A
MEANINGFUL
WORLD

How the Arts and Sciences Reveal the Genius of Nature

Benjamin Wiker & Jonathan Witt

IVP Academic

An imprint of InterVarsity Press
Downers Grove, Illinois

InterVarsity Press
P.O. Box 1400, Downers Grove, IL 60515-1426
World Wide Web: www.ivpress.com
E-mail: email@ivpress.com

InterVarsity Press® is the book-publishing division of InterVarsity Christian Fellowship/USA®, a movement of students and faculty active on campus at hundreds of universities, colleges and schools of nursing in the United States of America, and a member movement of the International Fellowship of Evangelical Students. For information about local and regional activities, write Public Relations Dept., InterVarsity Christian Fellowship/USA, 6400 Schroeder Rd., P.O. Box 7895, Madison, WI 53707-7895, or visit the IVCF website at <www.intervarsity.org>.

Scripture quotations, unless otherwise noted, are from the New Revised Standard Version of the Bible, *copyright 1989 by the Division of Christian Education of the National Council of the Churches of Christ in the USA. Used by permission. All rights reserved.*

Design: Cindy Kiple

Images: earth globe: Eyewire
galaxy: Stock Trek/Getty Images
red gerbera daisy: Stock disc/Getty Images
drawing by Leonardo da Vinci: Grafissimo/iStockphoto

ISBN 978-0-8308-2799-2

Printed in the United States of America ∞

Library of Congress Cataloging-in-Publication Data

Wiker, Benjamin, 1960-
  A meaningful world: how the arts and sciences reveal the genius of
  nature/Benjamin Wiker and Jonathan Witt.
    p. cm.
  ISBN-13: 978-0-8308-2799-2 (pbk.: alk. paper)
  ISBN-10: 0-8308-2799-4 (pbk.: alk. paper)
  1. Nature—Religious aspects—Christianity. 2. Creation 3.
  Meaning (Philosophy)—Religious aspects—Christianity. I. Witt,
  Jonathan James. II. Title.
  BT695.5.W565 2006
  231.7—dc22

2006013022

P	22	21	20	19	18	17	16	15	14	13	12	11
Y	33	32	31	30	29	28	27	26	25	24	23	22

In loving memory of

Francis Joseph Wiker, October 3, 2005

CONTENTS

ACKNOWLEDGMENTS

Many colleagues provided us valuable input and encouragement: above all, Mark Ryland, Guillermo Gonzalez and Jay Richards.

I'd (Benjamin) also like to thank my patient students who helped work through much of the material in this book long before it ever appeared in print: Charles Baran, Elizabeth Bedard, Marcus Bell, Bethany Carreon, Dallas Carter, Peter Cram, Carrie Cusick, Cefa Dasaret, Jeremy Dean, Matt Fenter, Michael Fink, Richard Gagne, Michael Jones, Anthony Luczak, Therese Naaden, Thomas Ostrowski, Will Sleever and Thomas Yager. I am also indebted to the Discovery Institute not only for funding my work but for discovering my indefatigable coauthor, Jonathan Witt, without whom this project would have been impossible and far less enjoyable. Above all, I thank my wife, Teresa, for her loving patience and support.

I'd (Jonathan) like to thank Benjamin Wiker for asking me into this project, and the editors of *Touchstone* magazine, who were adventurous enough to accept an odd mutt of a paper I sent them two years ago about the intersection of historical biology and aesthetics. Also, Stephen Meyer and the fellows of the Center for Science & Culture, for thinking we can win. Most of all, I am indebted to my wife, Amanda, first editor and encourager par excellence.

PROLOGUE

Crass Casualty obstructs the sun and rain.

Thomas Hardy, "Hap"

IMAGINE AN ALIEN WHO VISITS EARTH ONCE EVERY thousand years, a sort of intergalactic park ranger making the rounds, filing reports, just doing his job. Call him O. When he arrives at the turn of the third millennium, O is not surprised to find murder, envy, strife, wars and rumors of wars, vanity and vexation of spirit. His previous visits have led him to expect this of the human race. However, something does puzzle him. Although he finds that humanity has stumbled on the powerful investigative method we call science and, therefore, is experiencing unparalleled prosperity, the intellectuals in the very societies responsible for this happy revolution have adopted a posture of despair.

Seeking an explanation for this strange marriage of prosperity and despondency, O travels to the great universities of the West—but, alas, he is unable to arrange a single conversation with a professor. The intellectuals are either too busy writing carefully worded papers about the epistemological crisis or are so intent on seizing O as "exhibit A" in a fresh defense of cultural relativism that conversation is impossible.

Scratching his head, our bemused extraterrestrial moves on to the neighborhoods of the poor, to the churches, to the shop owners going about their work. One dieting mother complains cheerfully about making ends meet. An old man reading his horoscope grouses about pop culture going "down the toilet." Most, however, aren't sure what the extraterrestrial means by "the loss of meaning" and don't particularly want to figure it out. They're too busy watching the Home Shopping Network, plotting their next fantasy football trade or playing gory video games. "If you want to talk philosophy," an old fellow tells him, "try one of them fancy coffee shops."

And so our extraterrestrial heads downtown and enters the first one he sees. There he finds a mixed crowd. A few of the patrons look intensely alive. One of them pores over Yeats, another Pascal; a third scratches the notes of a new song onto the back of a napkin. But for every one of these, there are three striking various poses of exquisite despair over steaming cups of gourmet coffee.

"Why the hopelessness, the sense that life is meaningless?" the alien asks the sullen ones.

They complain about this or that politician, a war here or there. O points out that whereas tyranny and war have always been around, thousands of people sitting around stylish coffee shops complaining about the loss of meaning is new. "I found such despair among a handful of wealthy Romans in my visit before last," he explains, "but even there it was the exception."

Finally, a self-published poet in Birkenstocks looks up from his Nietzsche. "Hey alien man, we're atoms in the void, get it? Dust in the wind."

Still perplexed, the alien rubs his right head. "How did you arrive at that conclusion?"

The belated beatnik sets down his book. "Science. It's all been laid out in mathematical equations—survival of the fittest, everything's relative, indeterminable. You with me? Einstein, quantum mechanics, evolution."

"Scientists said it, you believe it, so that settles it?"

"Not the beaker jockeys. I'm talking about the visionaries—Dawkins, Sagan, Weinberg."

"And how did they arrive at their conclusions?"

"Why are you asking me? Do I look like a scientist? I'm just a fellow traveler, journaling the ride for the common man, tearing back the veil, the sheltering sky."

At last O has his answer. Somehow a group of influential human intellectuals have concluded that science has proven that the universe is essentially meaningless. Armed with this insight, he makes further inquiries of the other dispirited people in the coffee shop. They throw out a few catch phrases—"nature, red in tooth and claw," "the iron chain of causality," "survival of the fittest," "might makes right"—but none of them can explain in any detail how science has proven meaninglessness. Finally, an iron-haired religion professor looks over the top of his reading glasses and sighs, his patience all but spent. "It isn't about 'proof' anymore," he explains, speaking slowly and distinctly. "Rather, each of us must construct our own provisional 'meaning,' each our own 'truth,' and then dwell in it 'authentically.' " Around the room, heads nod in agreement.

O dutifully records all of this and returns to his ship and his home on the other side of the galaxy. There he tells his people about the strange despair infecting Western civilization in the wake of unparalleled prosperity.

His alien coworkers, however, insist he is pulling their legs. In desperation, he shows them holographic PowerPoint slides of sullen graduate students moping about their Ivy League campuses, soulless modern architecture heaped next to stone buildings older and infinitely more beautiful, book titles testifying to the death of meaning, and wealthy fashion models half-starved and aping death with charcoal makeup. Only then do his colleagues see that O speaks the truth. He reads to them from Stephen Crane, E. M. Forster—all of the modern purveyors of literary despair. The talk concludes with an excerpt from "Hap" by Thomas Hardy:

> If but some vengeful god would call to me
> From up the sky, and laugh . . .
> Then I would bear it . . .
> But not so. . . .
> —Crass Casualty obstructs the sun and rain.[1]

O's coworkers are baffled, groping to understand the source of the despair.

"A poison has entered human culture," O explains. "It's the assumption that science has proven that the universe is without purpose, without meaning—proven it so clearly that one need not even produce an argument."

The extraterrestrials are, of course, fiction. The poison, however, is real. This book is written as an antidote.

---

[1] Thomas Hardy, "Hap," in *The Norton Anthology of Poetry*, 3rd ed., ed. Alexander W. Allison (New York: W. W. Norton, 1983), pp. 494-95.

1

MEANING-FULLNESS
AND MEANINGLESSNESS

. . . what a falling-off was there!

Hamlet

THIS BOOK'S CENTRAL CLAIM IS SIMPLY STATED: the universe is meaning-full. It is rather curious that anyone would have to make an argument to that effect. But some today take it as a given that the universe is ultimately meaningless. Nihilism has spread from the philosophy of the "knowing" elite to the philosophy of everyday life. In the first half of the nineteenth century, philosopher Arthur Schopenhauer asserted against the rationalism of his day that nature is neither rational nor purposeful, that there is no benevolent God behind it and no goal after which it strives. Rather, nature is blind, a ruthless wilderness of struggle and destruction, and the world is without meaning. The root meaning of *nihil*—nothing—made *nihilism* a fitting name for such a dismal view of the cosmos.

But nihilism reaches much further back in Western culture. It is rooted in the modern acceptance of philosophical materialism, a view of the cosmos that can be traced back to the Greek atomists and, in particular, the philosopher Epicurus (341-270 B.C.).[1]

Building on the ideas of Leucippus and Democritus, Epicurus argued that the gods did not make the universe. The universe itself is eternal, and

---

[1] We say this knowing that Epicurean atomism—which formed the basis of Epicurus's atheistic, reductionist, materialistic system—did contribute to the modern focus on the atom, even while causing much intellectual mischief. In chapter five we will explore the complex and fascinating web of influences that led early chemists to uncover both the periodic table of elements and the atomic model. As we will show in several of the chapters that follow, many modern scientific discoveries powerfully challenge the metaphysical baggage of Epicureanism that has so powerfully influenced modernity.

in it, the ceaseless, purposeless swirl of mindless atoms create, quite by accident, everything from planets to stars, plants to people. Epicureanism was rediscovered in the early Renaissance and from there passed into the modern reductionist, materialist framework for describing nature—a program bent not only on denying anything beyond the material realm, but also on reducing everything to an ever smaller set of constituent parts and emphasizing these parts as ultimate reality. Epicureanism provided the prototype of the meaningless universe—godless, governed by chance, purposeless. Nihilism is its heir.[2]

But for centuries, materialism had a credibility problem. How could blind chance fashion something as intricate as an orchid or a butterfly? How could it create life? Half way through the nineteenth century, Charles Darwin suggested an answer. He argued that chance variation, coupled with the ruthless struggle for self-preservation, was the true creative force in the biological world. Occasionally, a random variation would occur that benefited survival and reproduction. This variation was more likely to get passed on to future generations than neutral or harmful variations. In this way, beneficial variations would pass on and accumulate and, over millions of years, lead to new species, orders and phyla.

Some found ways to hold Darwin's story in one hand and divinity in the other. But for many who accepted Darwin's newly fashioned creation story, all the glories of creation, all of its meaning and purpose, suddenly vanished. Human beings themselves became just one more accident of evolution, and a quite unfortunate one at that, for they alone of all the animals had been given the regrettable ability to grasp the ultimate meaninglessness of their own existence.

At the end of the nineteenth century, philosopher Friedrich Nietzsche provided the West with the most ruthless account of the meaningless universe. Taking both Schopenhauer and Darwin to heart, Nietzsche proudly declared that all philosophy, all religion, all science, all literature, all art were only so many desperate attempts to paint meaning on a meaningless cosmic canvas. We must "recognize untruth as a condition of life," he explained, and all attempts to portray truth are merely fictions masking the will to power.[3]

---

[2]For a thorough account of the pedigree of materialism, see Benjamin Wiker, *Moral Darwinism: How We Became Hedonists* (Downers Grove, Ill.: InterVarsity Press, 2002).
[3]Friedrich Nietzsche, *Beyond Good and Evil,* trans. Walter Kaufmann (New York: Vintage, 1966), sec. 4.

Nietzsche's own attempts to create meaning upon ultimate meaningless-ness ended in insanity. Nevertheless, by the twentieth century, nihilism was fashionable. Philosopher Jean-Paul Sartre offered a literary account in *Nau-sea*, a novel about the utter meaninglessness of human existence with a thinly disguised Sartre as the lead character, Roquentin. Here was the truly philosophical man who alone recognized the hideous nature of reality, hideous enough to bring on a kind of metaphysical sickness (hence the title).

Later the theme found its way into a variety of brooding plays, poems, stories and films and eventually begat a popular disciple in Woody Allen, who offered moviegoers a friendlier, self-congratulatory, domesticated nihilism-as-autobiography. Finally, at the end of the century, nihilism entered the mainstream, taking its place at the decentered center of the 1990's most popular sitcom, *Seinfeld,* a show, as its creators proudly proclaim, about nothing. In one episode the characters actually get lost in a mall's parking garage—appropriate to *Seinfeld*, since it's a world in which the search for meaning and value is less a valiant quest than a sort of aimless browsing at the local mall.[4]

As Karen Carr noted in 1992, "Nihilism, the bane of the nineteenth century, is fast becoming the banality of the late twentieth century."[5] As a banality, the assumption of meaninglessness has become commonplace, an obvious beginning point that, ironically, has become a truism. It is the last truth one can still assert in the company of intellectuals without embarrassment, having now the status of a conversational icebreaker, a cocktail party talking point that has taken the place of the weather.

But for all the respect it commands, the assumption of meaninglessness *is* only an assumption, a dogma that keeps many from seeing what should be obvious: the universe, rather than being devoid of meaning, is, like a great work of art, full to overflowing with meaning: complex, integrated and intelligible order, rather than senseless piles of gibberish. Contra nihilism, the universe has more meaning than we can imagine—layers of meaning that scientists, among others, continue to discover. It's a point we'll come back to again and again. Our argument is borne along, not in spite of recent scientific discoveries, but by those discoveries.

We now know, for instance, that even the simplest cells are full to overflowing with meaningful form. Thanks to discoveries in the second half of

---

[4]See Thomas H. Hibbs, *Shows About Nothing: Nihilism in Popular Culture from* The Exorcist *to* Seinfeld (Dallas: Spence Publishing, 1999).

[5]Karen L. Carr, *The Banalization of Nihilism* (New York: SUNY Press, 1992), p. 140.

the twentieth century, we understand that that the cell includes an alphabet and an extraordinary amount of functional genetic text. As software giant Bill Gates explains, "DNA is like a computer program but far, far more advanced than any software ever created."[6] DNA involves a four-letter alphabet written onto the double helix inside living cells. These four letters are used to construct the larger alphabet of 20 amino acids used to build proteins.

If only a little code were needed for the simplest self-reproducing cell—say a dozen amino acids of the right type and order—chance would do nicely as an explanation. But in even the simplest cells, the more than 30,000 different kinds of protein chains—each tailored to a particular task—are often hundreds of letters long. And even that sophisticated genetic software isn't enough for life. These genes can only function within the exquisitely ordered living structure, the cell. As New Zealand geneticist Michael Denton explains,

> Although the tiniest bacterial cells are incredibly small, weighing less than 10^{-12} gms, each is in effect a veritable micro-miniaturized factory containing thousands of exquisitely designed pieces of intricate molecular machinery, made up altogether of one hundred thousand million atoms, far more complicated than any machinery built by man and absolutely without parallel in the non-living world.[7]

A minimally functional cell would contain at least 250 genes and their corresponding proteins.[8] The odds of a primordial soup randomly burping

---

[6]Bill Gates, *The Road Ahead* (Boulder, Colo.: Blue Penguin, 1996), p. 228.

[7]Michael Denton, *Evolution: A Theory in Crisis* (Chevy Chase, Md.: Adler & Adler, 1986), p. 250.

[8]See Stephen Meyer, "DNA and the Origin of Life: Information, Specification and Explanation," in *Darwinism, Design and Public Education,* ed. John Angus Campbell and Stephen C. Meyer (East Lansing: Michigan State University Press, 2004), pp. 223-85. The essay engages the ongoing conversation among molecular biologists and information theorists, drawing on works too numerous to list here. A few that are particularly relevant to this aspect of his discussion are J. Reidhaar-Olson and R. Sauer, "Functionally Acceptable Solutions in Two Alpha-Helical Regions of Lambda Repressor," *Proteins, Structure, Function and Genetics* 7 (1990): 306-10; D. D. Axe, "Biological Function Places Unexpectedly Tight Constraints on Protein Sequences," *Journal of Molecular Biology* 301, no. 3 (2000): 585-96; Michael Behe, "Experimental Support for Regarding Functional Classes of Proteins to Be Highly Isolated from Each Other," in *Darwinism: Science or Philosophy?* ed. J. Buell and V. Hearn (Richardson, Tex.: Foundation for Thought and Ethics, 1994), pp. 60-71; H. P. Yockey, *Information Theory and Molecular Biology* (Cambridge: Cambridge University Press, 1992), pp. 246-58; Elizabeth Pennisi, "Seeking Life's Bare Genetic Necessities," *Science* 272, no. 5265 (1996): 1098-99; A. Mushegian and E. Koonin, "A Minimal Gene Set for Cellular Life Derived by Comparison of Complete Bacterial Genomes," *Proceedings of the National Academy of Sciences, USA* 93 (1996): 10268-73; and C. Bult et al., "Complete Genome Sequence of the Methanogenic Archaeon, *Methanococcus jannasch,*" *Science* 273 (1996): 1058-72.

up a concoction even half its length are vastly lower than one chance in 10^{150}. The universe isn't big enough, fast enough or old enough to generate the probabilistic resources to tame an improbability that large.[9] Even if, against all odds, a primordial soup had burped up the necessary set of protein chains, then, as we'll show in chapter eight, this still wouldn't be enough. The functioning cell isn't a set of requisite parts existing in an indiscriminate pile, or even a neatly ordered series of piles, any more than all the parts lined up neatly on shelves at your local car parts store are a car. As we discuss at greater length in chapter eight, the parts must be ordered in regard to space and time before they can function in the intricate life of a living cell. Genetic information, to function as information, must exist in the living structure within which it has meaning.

At least one philosophical materialist, eminent British philosopher Antony Flew, revisited the origin-of-life problem recently and began to rethink his position. Flew has long been held up by atheists and skeptical scientists as a beacon of reason and as a model of incisive, rigorous and enlightened thinking about the question of design. For many years he argued that there should be a presumption of atheism when approaching the question of origins, and for many years he insisted there was nothing in nature to overcome that presumption. But as he explained recently, he now has joined design theorists in seeing something more meaningful than chance or lawlike activity in the origin of life,[10] in seeing something like the fingerprints of an Author. He was not misled but led to that conclusion. As he explained, he "had to go where the evidence leads," and the evidence led him to reject the chipper atheism characteristic of leading Darwinist Richard Dawkins:

> It seems to me that Richard Dawkins constantly overlooks the fact that Darwin himself, in the fourteenth chapter of *The Origin of Species*, pointed out that his whole argument began with a being which already possessed reproductive powers. This is the creature the evolution of which a truly comprehensive theory of evolution must give some account. Darwin himself was well aware that he had not produced such an account. It now seems to me that the findings of more than fifty years of DNA research have provided materials for a new and enormously powerful argument to design.[11]

---

[9]William Dembski, *The Design Revolution* (Downers Grove, Ill.: InterVarsity Press, 2004).
[10]Antony Flew, "My Pilgrimage from Atheism to Theism: An Exclusive Interview with Former British Atheist Professor Antony Flew" *Philosophia Christi* 6, no. 2 (2004) <http://www.biola.edu/antonyflew>.
[11]Ibid.

To anyone not educated out of such an intuition, this is a commonsense view hardly in need of an argument. The information in DNA is a lot like this book in your hands: it contains sentence after sentence in a very particular order, and those sentences are functional because they have pages on which to rest, pages made of the appropriate material, properly ordered and bound—that is, the exquisite architecture of the cell that makes possible the work of DNA. The discovery of the cell's riches is just one instance of what occurred so often in the twentieth century: scientists again and again found a depth of meaning, of intricate, intelligible order, in the natural world.

In exploring this meaning we will move forward by philosophical, literary, mathematical and scientific analysis. This broad array of intellectual tools is needed for two reasons. First and foremost, since the universe is full of meaning—so rich to overflowing with evidence of its ingenuity—a number of disciplines are needed to capture this superabundance. Second, we are trying to break a spell, a kind of intellectual blindness caused by the ingrained habits of dogmatic materialism, and that blindness has infected virtually every intellectual discipline.

A first step in breaking the spell is to investigate what exactly is meant by the term *meaningless*. The most obvious sense is associated with words, with speech as an act of communication. When we think of meaningless speech we have in mind gibberish. Unlike the speech of an impenetrable foreign language, gibberish is not merely unknowable information. In gibberish, there is no information to know.

Gibberish may result from an unintelligent cause, like a monkey banging out *ffvvvvvvuffffffffffffffff* while attempting to subdue a typewriter. Or a monsoon might scatter a game of Scrabble, creating a random, meaningless pile of letters on the ground. In both cases, our recognition that the string of letters is meaningless is both complex and commonsensical. Experience teaches us that neither a monsoon nor a monkey is much given to the creation of meaningful phrases. Further, even if we knew nothing of how the arrangement of letters came to be, we wouldn't assume they were words in a foreign language; the excessive repetition in the former case and the alinear arrangement of the Scrabble letters in the latter would preclude such an assumption.

Gibberish might also occur if a person intended to be meaningless—if, for instance, a child wrote *AijbqW/Waa!Oktnvaegawwmtd* and pretended it was a secret code. Strange, yes, but are we sure it's meaningless? Perhaps the kid isn't kidding. Or perhaps it's a phrase written in an ancient tongue known only to a handful of obscure linguists. Or maybe it is the private language of Wiker and Witt, one we use to send encrypted missives to each another over the Internet. How could one possibly know for certain that *AijhqW/Waa!Oktnvaegauwmtd* was a random, meaningless string of letters? One would need to be omniscient to rule out the possibility of semantic meaning.

Notice, however, that such an omniscient condition is merely an extension par excellence of the very thing that led us to infer meaninglessness. As readers of this book, you know the English language and you understood the context of the original *AijhqW/Waa!Oktnvaegauwmtd*. That context is a book written in English prose and, more specifically, a meaningful, English paragraph calling for (as an example) the inclusion of a meaningless string of letters. Understanding these things leaves one with a sense that the letters probably were semantically meaningless (though not without purpose). Omniscient knowledge of all languages everywhere combined with an intimate and complete knowledge of the two authors of this book would merely cinch what you already understand to be the best explanation.

While we poor mortals are not omniscient, we can understand even in our own case that the detection of meaninglessness depends on our understanding of preexistent meaning, on our knowledge of language (most immediately, in this case, the English language surrounding the gibberish) and of the particular English meanings leading up to the string of nonsense letters. In other words, the recognition of disorder depends on order, so that without meaning, the notion of gibberish (ironically) wouldn't mean anything. This is evident in the very term *meaningless,* a subtraction or falling off from meaning. Meaninglessness is, therefore, what we might call parasitic or, with less of an edge, entirely dependent on preexisting meaning to have meaning.

But even meaning itself is dependent. Meaning doesn't stand alone; it points to something else. A crucial implication of this is that meaning isn't just a matter of letters accumulating into words. The backbone of language is the noun, and nouns are about *things.*[12] Their significance is that they signify; with the other parts of speech they are made to speak about reality, to say something about something. And so the following is also gibberish: "Un-

---

[12]Of course, nouns are regularly and rightly stretched to name all kinds of things that aren't things in any ordinary sense of the word (things like *incoherence*), but these ultimately depend on direct connections to things that are (e.g., *coherence*).

der crying the of trundle breakfast paladin glint." The string of words doesn't have meaning, even though the individual words do, because the string of words is not governed by an overarching unity or pattern that refers to something. A sign of this is that it has no defined subject and predicate, preventing it from saying something about something.

Indeed, a certain amount of gibberish would persist if we strung together sentences that were utterly unconnected to each other. Absent an overarching meaning that unites the several sentences of a paragraph, the straggling line of sentences collapses into a mob, a jumble of disconnected parts.

Here again, we recognize incoherence because we understand what it means to be coherent. We also recognize that meaning has a definite top-down, whole-part direction. That is, the larger context of a paragraph provides a sentence with meaning; the larger context of a sentence provides meaning for its words; and the individual letters convey meaning only when they become part of words.

We develop that point further in the next chapter, but now let's move on to another sense of meaninglessness, one that differs from gibberish. In a second sense, something can be meaningless, not in itself, but *to us* because we haven't deciphered it yet. In this sense, something isn't meaningless in itself (as with the monkey's contribution to experiential poetry, *ffv-vvvvvuffffffffffffff*). Rather, it is only meaningless *to* the viewer or investigator. Its meaning is knowable but not yet known.

Thus, for example, when I (Benjamin) look under the hood of my 1989 Ford F-150, I see a lot of parts. It looks to me like a confused jumble of hoses, knobs, wires, boxes and springs. If I pick up a part (which I often do, since it regularly sheds them) and ask, "What *is* this?" (which I often do, since I am mechanically challenged), I have no answer. The part is meaningless to me. I don't know what it does. I don't even know where it came from.

When I consult the *Haynes Repair Manual for Ford Pick-ups and Broncos, 1980-1996*, I am thrown into even greater confusion and angst. I read the text. I know all the English words. But since I am terminally ignorant of the way an engine works, the repair manual is as meaningless as the jumble of parts under the hood. *To me*, that is. Happily, my mechanic understands both quite well and quite cheerfully extracts the cash from my wallet as a just penalty for my ignorance. If I took the time to apprentice myself to a mechanic instead of forever scribbling away at books, then the engine and manual both would become meaningful to the degree that I learned what an engine is and how the many parts serve to make the engine run. And as I progressed, I would become strongly convinced that every part, no matter how bizarre looking or seemingly inconsequential, had an intelligible function, even though I didn't know it yet.

Consider the words of physicist Steven Weinberg in the penultimate paragraph of *The First Three Minutes:*

> It is almost irresistible for humans to believe that we have some special relation to the universe, that human life is not just a more-or-less farcical outcome of a chain of accidents reaching back to the first three minutes [after the big bang], but that we were somehow built in from the beginning. As I write this I happen to be in an airplane at 30,000 feet, flying over Wyoming en route home from San Francisco to Boston. Below, the earth looks very soft and comfortable—fluffy clouds here and there, snow turning pink as the sun sets, roads stretching straight across the country from one town to another. It is very hard to realize that this all is just a tiny part of an overwhelmingly hostile universe. It is even harder to realize that this present universe has evolved from an unspeakably unfamiliar condition, and faces a future extinction of endless cold or intolerable heat. The more the universe seems comprehensible, the more it also seems pointless.[13]

On Weinberg's view, the retail meaning of daily human commerce is only superficially meaningful, for the ostensible pointlessness of the beginning robs any of the busy goings-on of the human beings 30,000 feet below him of any ultimate meaning. *Because* "human life is . . . just a more-or-less farcical outcome of a chain of accidents reaching back to the first three minutes" (that is, the more we understand scientifically about the universe), *then* the "more it also seems pointless." Weinberg is claiming that, *as a matter of scientific fact,* we have demonstrated that the universe is meaning*less,* a purposeless whirl of matter, devoid of intelligent governance by a benevolent, divine being.

We should now recognize Weinberg's view as encapsulating precisely the view of meaninglessness that undergirds both philosophical and banal cultural nihilism. In his belief that the universe is ultimately a meaning*less,* purposeless whirl of matter, devoid of intelligent governance by a benevolent, divine being, he is the intellectual heir of Epicurus, Schopenhauer and Nietzsche, and he helps to provide the intellectual, scientific "verification"

---

[13]Steven Weinberg. *The First Three Minutes: A Modern View of the Origin of the Universe* (New York: Basic, 1977), p. 154.

of the popularized nihilism of Sartre, Woody Allen and even *Seinfeld.*

As a Nobel Prize-winning physicist, Weinberg certainly deserves a hearing for his position. But more fundamental than what he says is how he *acts* as a scientist. Does he behave like someone who believes the world is intrinsically meaningless?

Can a scientist be a scientist if the universe is ultimately meaningless? Science itself, by its very activity, reveals at least one level of meaning in the universe: intelligibility is found in the nature of things themselves, in the regular form and order of things accessible to scientific investigation, even in regard to our understanding of chance. To grasp this, imagine what it would be like to live in a truly disordered world, a reality of gibberish where there was no continuity between events and things just blinked in and out of existence, changing from one shape to the next in completely unpredictable ways. In such a universe, every event would be meaningless against a backdrop of total disorder. Events could be neither connected nor disconnected, no patterns could be discerned and, hence, there could be no science (or life, of course).[14] From this, it is clear that order makes science possible: a context of order allows things to be known and connected and allows us to make sense of sensation.

Most scientists accept this level of meaning without considering the trouble it might cause for the notion of wholesale pointlessness in the universe. Indeed, Weinberg is sufficiently myopic about this inconsistency that he can argue for wholesale pointlessness in his penultimate paragraph and, apparently without even realizing it, undercut himself in the subsequent, final paragraph:

> But if there is no solace in the fruits of our research, there is at least some consolation in the research itself. Men and women are not content to comfort themselves with tales of gods and giants, or to confine their thoughts to the daily affairs of life; they also build telescopes and satellites and accelerators, and sit at their desks for endless hours *working out the meaning of the data they gather.* The effort to understand the universe is one of the very few things that lifts human life a little above the level of farce, and gives it some of the grace of tragedy.[15]

---

[14]See the insightful essay by physicist Eugene Wigner, "The Role of Invariance Principles in Natural Philosophy," in *Symmetries and Reflections: Scientific Essays of Eugene P. Wigner* (Woodbridge, Conn.: Ox Bow Press, 1979), esp. pp. 28-29.

[15]Weinberg, *First Three Minutes,* pp. 154-55 (emphasis added).

Note that he does not say that men and women work for endless hours spinning out physicists' fantasies, but that they work out the *meaning* of real data. This assumes, of course, that the meaning is *in* the data or, to be more exact, that there is an intelligible order that the data reflect or convey, an order that, surprisingly, can be deciphered by human beings. Science is a meaningful activity precisely *because* the universe itself is meaningful and human beings have the strange capacity to understand it.

Recall the film *A Beautiful Mind,* in which future Nobel laureate and schizophrenic John Nash imagines he has been enlisted to intercept and decode enemy messages hidden in seemingly innocuous magazine and newspaper articles. Nobody hired Nash to do any such thing, and the supposedly hidden meanings he finds are only the invention of his disturbed mind. If we developed Weinberg's nihilistic conception of the universe into a logically consistent view of the world, this would describe all scientists, all mathematicians—madmen sunk in a mirage of meaning amidst a desert of meaningless disorder. Watching the movie, we do not view Nash's sinking into madness as farce. We view it as tragedy, and Weinberg, undoubtedly, did as well. Rather, Nash's condition strikes us as tragic precisely because here before us is a human mind capable of discovering real meaning deep in the order of things, but instead it is cast down, a genius fallen into chaos.

Most scientists aren't mad, although many fashion for themselves a kind of split personality wherein they act as if the universe is knowable, as if it has intrinsic meaning that they are discovering in the data, even while they proclaim that the universe is ultimately meaningless. The view isn't quite self-evidently contradictory since they, no doubt, would draw a distinction between wholesale and retail meaning. But notice where the burden of the argument lies. It's much more natural to suspect that if we find retail meaning, then there is also wholesale meaning—even though we may not yet have fully deciphered it.

If such is the case, then mysteries we encounter in the natural world are actually in accord with the second sense of meaning and meaninglessness outlined above, wherein something is meaningful but as yet undeciphered. We do indeed know a lot about nature, and that we can know this much is the result of it having an intrinsic, intelligible order that we could discover. That we can know even more—that we have not yet plumbed the depths of nature's order, that we continue to uncover layers of surprising complexity and beauty, that the universe proves itself again and again to be ingeniously wrought in a way accessible to the scientific enterprise—is a great sign that nature is not pointless but meaning-full, a work not of chance but of genius.

This would be the evident conclusion of all practicing scientists were it not for the dogma of reductionist materialism, which insists that the material realm is all that exists and that ultimate reality is to be found in the smallest constituent parts we can find. The everyday experience of these men and women is one of searching out the intrinsically intelligible order of nature, but the materialist dogma barks out that blind forces are the ultimate cause of everything and that the order around us is unintended, pointless.

But for those who will listen, the order of nature bespeaks an altogether different lesson. It's whispering to us of the genius of nature, and we can hear it if we will but awaken from the spell of materialism.

To break that spell, we must look for those things on the stage of materialism that are out of place, train our eyes on what doesn't fit. Rather than leap right into an analysis of nature or even of scientists studying nature, we begin with an analysis of human genius itself, specifically the tragedy and comedy of William Shakespeare. We do this because the most poisonous effect of the materialist spell is the way it clouds our self-understanding. Materialist reductionism seeks to give an entirely material explanation of human intelligence, one that reduces it to a string of pointless material causes. It must kill the soul and, in the process, reduce all the evident genius of humanity to dust. To this end, some materialists have gone after Shakespeare with a tireless ingenuity. In the two chapters that follow, we will keep company with Shakespeare's genius and consider whether news of its dissolution has been greatly exaggerated.

We will show how the meaning of a literary work does not develop merely sequentially but in a recursive and goal-directed relationship with the work as a whole as it exists and takes shape in the author's mind. Living organisms, we argue, function in the same way, so they are not amenable to the blind, step-by-step process suggested by modern evolutionary theory. Our journey with Shakespeare concludes with a description of the elements of genius in Shakespeare's art, well positioning us to consider whether nature itself possesses not merely the hallmark of design but of genius.

That suspicion will lead us toward other demonstrations, other signs to guide us as well as other obstacles along the way. We will segue from human nature to nature, with the next stage in our analysis a kind of midway point—mathematics, one of the great intellectual tools that we human knowers use to investigate the order of nature, the very tool which scientists like Weinberg so successfully employ to uncover the meaning of the data. As will become clear, the glory of mathematics as a human art manifests the glories of the human intellect, and here too there is a genius to experience, that of the Greek mathematician Euclid. As his insights unfold into the most distant stretches of the cosmos, we ponder, as Albert Einstein did, the striking fact that the universe is comprehensible, that mathematics illuminates nature by mapping forms of order as small as the bonds within an atom and as broad as the universe. On materialist grounds, why should it? Why should there be any connection whatsoever between the highly abstract, formal relationships of numbers and figures and the order of nature? Why, in short, is nature amenable to mathematical analysis?

From a consideration of mathematics, we turn to the world of chemistry, to which materialists claim everything above physics can be reduced, from nature itself to the very activity of scientists who study nature. Against this materialist claim, we again offer the evidence of genius, studying the string of geniuses who discovered and assembled the periodic table of elements and arguing that a materialist account of the human mind fails to account for such extraordinary mental feats. Second, the periodic table itself is a masterpiece of order, precision and intellectual beauty, an order that appears designed for both life and discovery. Chemistry, rather than being that which everything can be reduced to, seems to point upward toward life, as if meticulously crafted for just such a purpose. Finally, in the way the order of the elements has wrung from great scientific minds their best efforts and only then yielded up its secrets, the genius of the elements would seem also to have had in mind not merely life, but specifically the development of human genius.

The suggestion is scandalous only in the narrow room of materialism. Outside that room, we will argue, the conclusion is as natural as a leaf reaching for sunlight. Against the materialist attempts to reduce biology to chemistry, we find instead that the latest science is uncovering more and more evidence that the elements are strangely fit for biology, the lifeless fashioned for the living.

Following this flow, we turn our attention to biology last of all, to the ingenuity of living things. Here we intend to show that it is as misguided to reduce the living world to blind forces as it was to reduce Shakespeare to brute urges. The layer upon layer of complexity and beauty, the ingenious designs and the majestic integration all point beyond pointlessness to a meaningful world and, more than this, to the genius of nature.

Contemporary design theorists until now have not explored the category of genius in a formal and extensive way but instead have focused on one narrow quality of intelligence: its ability to choose among options for a future end. Perhaps some have felt that rationally reconstructing an argument to genius would be difficult to do without creating vulnerabilities. After all, even a poorly constructed automobile possesses the hallmark of design. In the same way, a species with an apparent defect could nevertheless possess the clear signature of design, even if the reason for the defect remained unexplained.

But we will argue that the explanatory category of genius actually strengthens rather than weakens the design argument. As philosopher Jay Richards and astrobiologist Guillermo Gonzalez note in *The Privileged Planet,* where one adds to evidence of design evidence of a specific purpose, the argument for design becomes more robust.[16] We would like to push the argument further, demonstrating that where one can show evidence not only of design and purpose, but of genius, the argument to design becomes still more robust, providing a tighter specification as well as additional explanatory power.

But what about seemingly bad design? Why is the universe largely uninhabited and uninhabitable? Why are there earthquakes, back aches, junk DNA? It's important to make clear at the outset that seeing the signature of genius in nature does not entail a Panglossian view of nature as "the best of all possible worlds." One can recognize the hand of genius without turning a blind eye to disease and deformity, pain and suffering.

One of us hopes to address the question of apparent bad design at greater length in a future book.[17] As the reader will see, our experience with the artistic works of human genius lead us to expect depths that outstrip our immediate understanding, even in regard to our judgments concerning the caliber of particular aspects of the design. Moreover, the careful study of works of genius encourages investigative optimism—the conviction, born of experience, that many inscrutable things in a work of genius are only apparently so and, that, with sustained effort, the work will yield up more and more of

---

[16]Guillermo Gonzalez and Jay Richards, *The Privileged Planet: How Our Place in the Cosmos Is Designed for Discovery* (New York: Regnery, 2004), p. 307.

[17]The book is in progress with the provisional title *Darwin and Shakespeare: Aesthetics and Bad-Design Arguments Against Intelligent Design.* It builds on Jonathan Witt, "The Gods Must Be Tidy! Is the Cosmos a Work of Poor Engineering or the Gift of an Artistic Designer?" *Touchstone,* July/August 2004, pp. 25-30.

its secrets, giving us a clearer eye and a more comprehensive viewpoint. In sum, the scientist who recognizes nature as a work of genius can explain—rather than merely explain away—our collective experience of repeatedly uncovering new mysteries and of repeatedly uncovering answers to those mysteries.

And what about the problem not merely of bad design but of evil? That subject isn't the focus of this book, but our argument lends support to an orthodox view of evil as parasitic on good, in the same way that disorder is parasitic on order and meaninglessness on meaning. To view evil in this way does not eliminate evil or even render it a thing of little account; but it does put it in proper perspective. In some fundamental way, goodness, order and meaning are deeply related, and evil, disorder and meaninglessness are related in that each is a falling off from being. We shall return to this insight in the last chapter, but only after an extended examination of the many ways the world is charged with meaning.

2

HAMLET AND THE
SEARCH FOR MEANING

It is through wonder that human beings [anthrōpoi]
both now and at first began to philosophize.

Aristotle, *Metaphysics*

CHARLES DARWIN AND WILLIAM SHAKESPEARE are unlikely bedfellows, so much so that to treat them together in an academic work cuts across the grain of our Enlightenment-inscribed intuitions. But that is part and parcel of our thesis. Shakespeare's works have a rich, irreducible complexity that doesn't fit into the reductive, materialist world of Darwin.

Many since Darwin have attempted to reduce Shakespeare's gifts to blind material causes like sexual selection, but a journey through Shakespeare's most celebrated work, *Hamlet,* shows that all such attempts fail. The overarching form of Shakespeare's works and the intricacy of his craft demonstrate not only the obvious (that an intelligent cause of high order is responsible for these plays), but that this cause cannot be reduced to blind mechanism. In light of this, we might begin to wonder whether not just Shakespeare, not just human art, but nature itself has been unjustly, and irrationally, reduced to far less than it is.

As literary critics were just beginning to try to force the beauties of Shakespeare into the box of materialism, scientists intent on defending Darwin's theory were also sniffing around the Bard, hoping to tame him for their purposes. They happened upon an arresting image that took a variety of forms but went something like this: if a million monkeys banged away on typewriters for a million years, eventually they would generate the entire works of Shakespeare. This claim has cast a kind of malignant charm for far too long. Thus, it wasn't until 2002 that enterprising researchers finally set about to test the performance levels of typing monkeys at Plymouth University in England. The researchers left a computer in a cage with six Sulawesi crested macaques at Paignton Zoo in southwest England for a month. The literary results were, if nothing else, original. The work begins thus:

ff

vvvvvvvpppssgggggggggggggggggggggggggggggggggggggggggggggggggggggggg
gggggggggggggggggggggggggggggggggggggggggggggggggggggggggggggggggggg
gggggggggggggggggggggggggggggggggggggggggggggggggggggggggggggggggggg
ggggggggggggggggggggggggggggggggggggggggggggggggggggggggggggggggggggg
gggggggggggggggggggggggggggggggggggggggggggggggggggggggggggggggggggggg
gggggggggggggggggggggggggggggggggggggggggggggggggggggggggggggggggggggg
gggggggggggggggggggggsssssssssssssssssssssssssssssssssssssssssssssssss
sssssssssssssssssssssssssssssssssssssssssssssssssssssssssssssssssssss
sssssssssssssssssssssssssssssssssssssssssssssssssssssssssssssssss55555S
ssssssssssssssssssssssssssssssssssssssssssssssssssssssssssssssssssssss
ssssssssssssssssssssssssssssssssssssssssssssssssssssssssssssssssssssss
ssssssssssssssssssssssssssssssssssssssssssssssssssssssssssssssssssssss
ssssssssssssssssssssssssssssssssssssssssssssssssssssssssssssssssssssss
ssssssssssssssssssssssssssssssssssssssssssssssssssssssssssssssssssssss
ssssssssssssssssssssssssssssssssssssssssssssssssssssssssssssssssssssss
ssssssssssssssssssssssssssssssssssssssssssssssssssssssssssssssssssssss
ssssssssssssssssssssssssssssssssssssssssssssssssssssssssssssssssssssss
ssssssssssssssssssssssssssssssssssssssssssssssssssssssssssssssssssssss
ssssssssssssssssssssssssssssssssssssssssssssssssssssssssssssssssssssss
ssssssssssssssssssssssssssssssssssssssssssssssssssssssssssssssssssssss
ssssssssssssssssssssssssssssssssssssssssssssssssssssssssssssssssssssss
ssssssssssssssssssssssssssssssssssssssssssssssssssssssssssssssssssssss
ssssssssssssssssssssssssssssssssssssssssssssssssssssssssssssssssssssss
ssssssssssssssssssssssssssssssssssssssssssssssssssssssssssssssssssssss
ssssssssssssssssssssssssssssssssssssssssssssssssssssssssssssssssssssss
ssssssssssssssssssssssssssssssssssssssssssssssssssssssssssssssssssssss
ssssssssssssssssssssssssssssssssssssssssssssssssssssssssssssssssssssss
ssssssssssssssssssssssssssssssssssssssssssssssssssssssssssssssssssssss
ssssssssssssssssssssssssssssssssssssssssssssssssssssssssssssssssssssss

ssssssssssssssssssssssssssssssssssssssssssssssssssssssssssssssssssssssssssssss
ssssssssssssssssssssssssssssssssssssssssssssssssssssssssssssssssssssssssssssss
ssssssssssssssssssssssssssssssssssssssssssssssssssssssssssssssssssssssssssssss
ssssssssssssssssssssssssssssssssssssssssssssssssssssssssssssssssssssssssssssss
ssssssssssssssssssssssssssssssssssssssssssssssssssssssssssssssssssssssssssssss
ssssssssssssssssssssssssssssssssssssssssssssssssssssssssssssssssssssssssssssss

With an eye keenly attuned to essentials, researcher Mike Phillips noted, "They pressed a lot of S's." As it turns out, the monkeys—Elmo, Gum, Heather, Holly, Mistletoe and Rowan—seemed less interested in leaving their marks on literature than in leaving their marks on the computer. Rather than sitting down immediately to type at the beck of the muse, "the lead male got a stone and started bashing the hell out of it," reported Phillips with candor. "Another thing they were interested in was defecating and urinating all over the keyboard."[1] Suffice it to say, their literary efforts fall a good deal short of the Bard.

Given the squishy empirical foundations of the Darwinists' typing monkeys claim, we may return to those questions that should have met the claim in the first place. For one, are monkeys the kind of natural thing that will stay bent over the keyboard, seized by a maniacal work ethic, methodically pecking away at various letter combinations for a million years? The answer, for anyone with the slightest familiarity with monkeys, is no. Monkeys are the kind of natural thing that will, first of all, give the keyboard a good drubbing to let it know where it stands in the pecking order, then treat it to several scatological initiations, and finally push down on one or two of the keys—but do so very occasionally (enough to produce only about five pages during an entire month).

With such a meager and unmethodical output, even the simplest English sentence of the "See Spot run" type is beyond the capacities of *real* monkeys, no matter how long you give them. The task would more than outstrip the lifespan of any single group of monkeys. Further, whatever time they had probably would not be spent unnaturally bent over a keyboard. If their month-long labor was any indication of their labor-to-leisure ratio, the

---

[1]For two media reports on the Shakespearean monkeys, see Brian Bernbaum, "Monkey Theory Proven Wrong," *CBS News,* May 9, 2003, and David Adam, "Give Six Monkeys a Computer, and What Do You Get? Certainly not the Bard," *The Guardian,* May 9, 2003. These are available online at the Access Research Network <http://www.arn.org/docs2/news/monkeysandtypewriters051103.htm>.

Paignton Zoo monkeys would spend a mere 0.00004 percent of their time typing; the rest of the time would be spent sleeping, scratching, eating, staring, playing, grooming and ensuring that future generations would exist to carry on the work of creating dramatic art of one form or another. It's worth adding that while familiarity might not breed contempt in monkeys, it does breed indifference to things inedible, so it is likely that, if left to themselves, their production would trail off in subsequent months. Finally, note that computers do not particularly appreciate being pounded, much less defecated on, and undoubtedly at some point would protest the treatment by ceasing to work.

Now it is easy to see why defenders of Darwin's theory would naturally be inclined to minimize the distance between the capacities of humans and other primates and, even more, why they would want to suggest that pointless physical processes like natural selection can surge forward, generation upon generation, incessantly plowing through variations and combinations with machinelike intensity, continually pushing lifeless chemicals upward toward a cathedral-like complexity. Connecting the monkeys to the alleged powers of natural selection also allowed Darwinism to rid itself of the embarrassment of riches so evident in the Bard's genius. If monkeys could knock out a Shakespearean tragedy given enough time, then what about creating Shakespeare himself? Couldn't he be almost as easily explained on Darwinian grounds?

It was a powerful rhetorical ploy, and one Darwinists have returned to. In recent times, Oxford zoologist and leading evolutionist Richard Dawkins updated the analogy, enshrining a particular phrase from *Hamlet* as demonstrative of the amazing powers of evolution to mimic not just intelligence, but the keen intellect of the Bard himself.

The phrase? "Methinks it is like a weasel" taken from act 3, scene 2 of *Hamlet*.[2] Dawkins uses it to demonstrate the difference between "single-step selection" and "cumulative selection" and to argue that, while the purely random generation of the entire phrase in one fell swoop (single-step selection) is dauntingly improbable (one chance in 3×10^{41}), we can quickly build to it given the powers of cumulative selection, wherein a procession of phrases, as evolutionary steps, ever more closely approximate the final

---

[2]Richard Dawkins, *The Blind Watchmaker: Why the Evidence of Evolution Reveals a Universe Without Design* (New York: W. W. Norton, 1996), pp. 45-50. The Shakespearean phrase comes from *Hamlet: A Norton Critical Edition* (New York: W. W. Norton, 1963), 3.2.348.

product, *Methinks it is like a weasel*. Dawkins is trying to demonstrate that in assessing the powers of chance to produce a living being or a complex organ, evolutionists are not claiming that, say, a functional grackle's wing pops into existence in one fell swoop. Rather, evolution always works by cumulative steps, building slowly to the goal through a long series of functional intermediates.

And so, according to Dawkins, if we allow the Shakespearean phrase to be generated cumulatively (using a computer, rather than a monkey), we move swiftly from a random starting point *(WDLMNLT DTJBKWIRZREZLM-QCO P)*, through various generations to the target phrase. In each generation, "mutations" allow random substitutions, thereby creating a pool from which the phrase most akin to *Methinks it is like a weasel* is selected as the foundation for the succeeding generation.

By such means we move from the starting nonsense string of letters to *WDLTMNLT DTJBSWIRZREZLMQCO P* in the second generation; after ten more generations we have *MDLDMNLS ITJISWHRZREZ MECS P;* after thirty, *METHINGS IT ISWLIKE B WECSEL;* and by the forty-third generation, *ME THINKS IT IS LIKE A WEASEL.* We are informed after this prodigious feat that "computers are a bit faster at this kind of thing than monkeys, but the difference really isn't significant."[3] And so the famous typing monkeys have been resurrected, and once again the artistry of nature and the Bard are well within the reach of blind evolution.

This computer's work is only an analogy, but Dawkins argues that it's a sound analogy, equating the powers of the computer's selection with the powers of natural selection. And so while nature has no long-term goal, or target phrase, each generation is governed by survival of the fittest, so that "either simple survival or, more generally, reproductive success" function as an immediate and ever present "target."[4]

There is one respect in which this computer analogy is dead on. Like the extraordinarily complicated "hardware" of the first self-reproducing cell needed before Darwinian evolution could begin its work (see chapter eight), the computer housing the sentence-generating program remains unexplained. After that the analogy falls apart. There are two nibbling problems with the analogy: first, unlike Darwinian selection, this computer knows precisely where it's headed; it isn't blind. It's headed for the target

---

[3]Dawkins, *Blind Watchmaker,* p. 49.
[4]Ibid., p. 50.

phrase already programmed into the computer. Thus, the program mimics guided or teleological evolution, not Darwinian evolution.

Second, the functional intermediates aren't functional. The original "sentence" Dawkins provides is not even functional; that is, *it has no meaning.* Indeed, his intermediate steps only begin to appear recognizably similar to the target phrase after about 75 percent of the generations have been produced. (This defect, albeit in a subtler form, bedevils the computer program Avida, an attempted simulation of Darwinian evolution reported in the journal *Nature* in 2003.)[5] But to mimic the alleged powers of evolution, Dawkins would have to transform one functional sentence into the target sentence through a series of functional, meaningful intermediates. The difficulty—the sort that plagues evolutionary accounts of significant genetic change—is that the space between markedly different functional sentences is astronomically large because functionality demands real words formed into the integrated whole of the larger order of a meaningful sentence, as governed by syntax, grammar, spelling and the accepted meaning of the individual words.[6]

If the computer program had been set up to properly mirror Darwinian evolution, all nonsense strings would simply be eliminated as gibberish,

---

[5]Richard E. Lenski, Charles Ofria, Robert T. Pennock and Christoph Adami reported results from the Avida computer simulation in "The Evolutionary Origin of Complex Features," *Nature* (May 8, 2003): 423:139-44. William Dembski critiques the simulation in the introduction to *Uncommon Dissent: Intellectuals Who Find Darwinism Unconvincing* (Wilmington, Del.: ISI Books, 2004). For an extended critique available online, see Royal Truman, "Evaluation of Neo-Darwinian Theory with Avida Simulations," International Society for Complexity, Information and Design (July 1, 2004) <http://www.iscid.org/boards/ubb-get_topic-f-10-t-000085.html>. As Truman explains, the simulation's central error is that the program grants "function" to intermediate steps with an almost flagrant disregard for biological realism: (1) "Miniscule genomes are assumed with far more superfluous than necessary genetic material"; (2) "Extremely high mutational rates are used," permitting "generation of logic functions before genome truncation would remove superfluous genetic material"; (3) "The physical machinery to transcribe, translate, and perform metabolic processes are not coded genetically and thus not subject to mutational damage"; (4) "Statistically insignificant sequence space distances are assumed between novel, more complex functions . . . an artifact of logic functions and not protein sequence space"; (5) "New, more complex functions, like EQU, once formed are virtually guaranteed to perpetuate . . . due to very high relative fitness rewards; and (6) "No path for graceful degradation of functions is provided . . . which permits 'natural selection' to easily identify degraded functionality with 100% accuracy." Another widely touted evolution simulation is T. D. Schneider, G. D. Stormo, L. Gold and A. Ehrenfeucht, "Information Content of Binding Sites on Nucleotide Sequences," *Journal of Molecular Biology* 188 (1986), 415-31. I. G. D. Strachan critiques it in "An Evaluation of 'Ev.'" International Society for Complexity, Information and Design (June 30, 2003) <http://www.iscid.org/papers/Strachan_EvEvaluation_062803.pdf>.

[6]See Michael Denton's incisive analysis of this point in *Evolution: A Theory in Crisis* (Bethesda, Md.: Adler & Adler, 1996).

only to be followed by another random go at the whole string on the next try. Even if the program had been designed to preserve individual English words as they occurred, it wouldn't have really mirrored the task Darwinian evolution has before it. As we will discuss in the final chapters, the jumps between functional genetic prose occur more at the level of sentences and even paragraphs.

What Dawkins's analogy lacks in rigor it makes up for with ingenious rhetoric. Dawkins could have chosen any brief quotation from *Hamlet* or from the entire corpus of Shakespeare. Why "Methinks it is like a weasel"? The reason is simple. Hamlet says these words while cloud gazing, and as Dawkins has famously stated, "Biology is the study of complicated things that give the *appearance* of having been designed for a purpose."[7] That is, a cloud shaped like a weasel appears to be designed, to be meaningful, but it isn't; likewise, living things like weasels appear to be the work of a great artist, but really they're just the accidental product of natural selection working on genetic variation.

So, the section in *The Blind Watchmaker* that adverts to Hamlet's phrase begins with a discussion of clouds that "through random kneading and carving of the winds, come to look like familiar objects." Dawkins then pauses to note, in a seemingly random way, that there is a "much published photograph, taken by the pilot of a small aeroplane, of what looks a bit like the face of Jesus, staring out of the sky." Such vaporous resemblances in both cases come about "by single-step selection," and hence appearances in clouds are not very close to reality.[8] However, since "cumulative selection" is so powerful in biology, "biological adaptations" nearly overwhelm us with the notion that the creatures before our eyes have been designed. Add to this the aside about Jesus in the clouds—a clever barb at Christians, implying as it does the nebulous origins of their beliefs—and we have before us an ingenious and meticulously structured piece of rhetoric.

Dawkins is even thoughtful enough to provide some much needed context for the "weasel" line, quoting the relevant passage that follows:

---

[7]Dawkins, *Blind Watchmaker*, p. 1 (emphasis added).

[8]It is difficult to generously reconstruct Dawkins's meaning when he describes cloud formation as single-step selection, akin to getting *Methinks it is like a weasel* in one randomly generated step. The generation of clouds is actually very complex, involving not only the various conditions that allow for liquid water on Earth and an atmosphere to hold it, but also the presence of the complex hydrological cycle, the heat of vaporization of water itself and the presence of cloud condensation nuclei from biological and nonbiological sources. Beyond this, it takes intelligence of some kind to imagine a similarity between a cloud and a weasel.

> *Hamlet.* Do you see yonder cloud that's almost in shape of a camel?
> *Polonius.* By th' mass, and 'tis like a camel indeed.
> *Hamlet.* Methinks it is like a weasel.
> *Polonius.* It is backed like a weasel.
> *Hamlet.* Or like a whale?
> *Polonius.* Very like a whale.[9]

It was quite sensible for Dawkins to give us this much context at the very least, but as we shall see, doing so undoes his argument. The isolated sentence *Methinks it is like a weasel* is a cipher. Only within a larger unity does it function properly. Of course, in an inconsequential way Shakespeare did generate it letter by letter when he actually wrote it down, in the same manner that the letters of this book are typed letter by letter. But Shakespeare created *Methinks it is like a weasel* in light of the dramatic unity of the whole play, and that is where it manifests its full meaning.

The necessity of referring to a larger context to give the target phrase sufficient meaning should have been clear to Dawkins himself, since he felt compelled to provide enough context to reveal the sentence's basic meaning (and indeed, a double meaning, given his desire to make the belief in design as foolish as assigning design to the particular shape of a cloud). The isolated sentence is a grammatical whole, but who is talking? Hamlet? Who is Hamlet, and what does he think is like a weasel? To whom is he speaking? Polonius? Who is Polonius? Why is Hamlet telling Polonius "it is like a weasel"? Why does he keep changing his mind? What's really at issue in the conversation? If we had never seen or read the play, we would be hard-pressed, given the sentence in isolation, to offer an account even of its barest meaning. Even with our first four questions answered, we're adrift.

This point is so important to our argument that we need to go into some detail about the passage, for it's only by ignoring the great depth of meaning found in the passage when studied in its larger context that one is able even to conceive of its being created by random processes. Not surprisingly (and as we will discuss in greater depth in a later chapter), Dawkins and other Darwinists treat biological traits and lines of genetic information in the same misguided way—as "phrases" that can exist somehow apart from the real, living dramatic unity of biological organisms.

So, then, what is the living, complex unity of which this Shakespearean phrase is an integral part? What is the dramatic unity for which this phrase

---

[9]*Hamlet* 3.2.346-51.

was generated? Briefly, Hamlet is a medieval prince of Denmark. His father (King Hamlet) has been secretly murdered by the king's own brother, Claudius, who then marries the old king's widow (Prince Hamlet's mother) a mere two months after the king's funeral. At the opening of the play, the young prince does not know of his uncle's unnatural treachery but is mourning his father's death and privately railing against his mother for marrying his uncle so swiftly. Hamlet learns of the true cause of his father's death from his father's ghost, news that understandably deepens Hamlet's inner turmoil. The prince's subsequent quest to avenge his father's death shapes the rest of the play and is the occasion for profound reflections on the nature of sin; human and divine punishment; love and loyalty; the divide between appearance and reality; the tangled web of providence, happenstance and free will; and even the nature of dramatic art itself as a means of bringing people to cathartic recognition of their evil crimes.

Of course, this short listing of its elements only touches the play's surface. More has been written about *Hamlet* than any other work of literature outside the Bible—testimony to its rich depths. Our task here is merely to hint at those depths and, more immediately, to convey just how dependent the weasel passage is on its larger context.

Shortly before Hamlet and Polonius have the interchange excerpted above, a play was performed in front of King Claudius and Queen Gertrude, a play chosen and modified by Hamlet to mirror Claudius's act of poisoning King Hamlet and marrying Queen Gertrude. As Prince Hamlet had hoped, the play goads Claudius into betraying clear signs of a guilty conscience, confirming for Hamlet that Claudius did indeed murder Hamlet's father. Rosencrantz and Guildenstern, courtiers and youthful friends of Hamlet, come to inform Hamlet that his mother now wishes to see him in private, but Hamlet can only treat them sardonically, for although they were his friends, they are now willing servants of Claudius (and in fact, have agreed to spy on Hamlet). Thus, events since his father's death have taught Hamlet that in regard to worldly ambitions and passions, mothers and friends sell their love and loyalty all too quickly.

After Hamlet accuses Rosencrantz and Guildenstern of trying to manipulate him like a musician's pipe for their own political gain, Polonius, Lord Chamberlain of the Court, enters to speak with young Hamlet. Once loyal to King Hamlet, Polonius is now absolutely committed to the usurping Claudius. He is the archetypal yes man who imagines he hunts "the trail of policy" with a keen and penetrating eye, when really he is trapped on the surface among a set of inadequate and reductive truisms. It's doubtful Polonius knows anything of the murder, but he only nods obediently when Claudius pops between young Hamlet and the throne, and he smiles approvingly when Claudius marries the old king's widow so quickly that, as Hamlet bitterly notes, the "funeral bak'd meats / Did coldly furnish forth the marriage tables."[10] Polonius is a court toady and Hamlet despises him, all the more because the man forbids his daughter, Ophelia, to have any private dealings with the prince.

That is a rough sketch of the context for the famous cloudy-weasel passage. The clouds go whichever way the wind blows and change shape with the shifting winds. So does Polonius. Hamlet lures him into agreeing that the cloud has first one shape, then another, then another. Hamlet is mocking Polonius's obsequiousness, which, in deference to royalty, will agree to anything said by a prince or king, no matter how absurd. Hamlet first remarks that the cloud looks like a camel, and Polonius swears "by th' mass" that it is "like a camel, indeed." It's instructive that Polonius would use such an oath—swearing by the most sacred sacrament of his church, wherein Jesus Christ is believed to be truly present, body, blood, soul and divinity, on the altar—in relation to such an ephemeral and trivial thing as the appearance of a cloud: it suggests the displacement of divine justice in the interest of political gain.

Hamlet, knowing Polonius's character all too well, then remarks of the cloud, "Methinks it is like a weasel," whereupon Polonius immediately changes his mind to conform to what he thinks Hamlet wants to hear, remarking that indeed "It is backed like a weasel." Of course, Hamlet means not only to expose Polonius's shallowness, but also to associate Polonius's character with that of the none-too-honorable animal itself. Finally, Hamlet remarks that the cloud really looks like a whale, and predictably, Polonius quickly agrees that it is "very like a whale." The context (Hamlet exposing Polonius as a yes man) clearly suggests that neither really thinks the cloud even looks particularly like a whale or a weasel, much less that it actually is one.

Polonius's problem isn't seeing design where it doesn't exist: his problem is failing to see design where it really is. King Hamlet's death wasn't happenstance; it was the result of Claudius's cold-blooded design. Polonius ignores Claudius's double motive for murder—the throne and the queen. He ignores the man's suspiciously aggressive grab for the throne over the head

---

[10]Ibid., 1.2.180-81.

of the popular and natural heir, Prince Hamlet, something he only managed because Hamlet happened to be away at the time. Polonius ignores Claudius's lusty grab for Queen Gertrude, ignores the message of the play within the play and ignores that the enraged Claudius clearly "doth protest too much" for an innocent king merely registering that the play is in bad taste and his poor nephew is suffering from acute paranoia. There is a pattern here. It isn't enough to convict Claudius, but it's more than enough to lead a reasonable man to suspect foul play. Quite irrationally, Polonius, as best we can tell, never does.

In short, because Polonius is a naive and obsequious weasel, he misses the signature of design where it does exist and claims to see a pattern in the clouds that does not exist. So the weasel passage really is about appearances, but not in the way Dawkins suggests. Dawkins isn't wrong to look for new significance in the text of *Hamlet*. Without pressing the distinction too far, it's sensible to distinguish between the meaning of a literary passage and its possible significance to this or that later situation.[11] But if one wishes to apply the passage to the distant realm of origins science, as Dawkins has attempted to do, a more natural application is this: some people will pretend to see things to suit their own purposes while missing the true signatures of design all around them, because to see the design and point it out would risk their position.

And as we return from significance to meaning, again it should be clear even from our short analysis that the weasel line takes its ripe and proper meaning, its very form, in large measure from the context of the entire play. The problem facing Dawkins, then, is that of generating not a single, relatively short phrase, but an entire play in which this phrase actually lives. One cannot begin with a randomly generated set of meaningless letters to the target phrase and then build randomly but cumulatively to the entire play in the way Dawkins suggests, because it is in light of the characters, plot and theme that the phrase itself has been created and has its functional, meaningful existence.

---

[11]The distinction is from E. D. Hirsch Jr., *Validity in Interpretation* (New Haven, Conn.: Yale University Press, 1967). The materialist attack on the Author of nature led to a loss of confidence in the literary author as source to stabilize meaning. In the latter half of the twentieth century, it was considered gauche to try to get at an author's intended meaning. The author is dead, Roland Barthes declared, and we manufacture our own meanings by one means or another. Hirsch's book is a rear-guard defense of authorial intention. Perhaps an even more rigorous and incisive defense is P. D. Juhl, *Interpretation: An Essay in the Philosophy of Literary Criticism* (Princeton, N.J.: Princeton University Press, 1980).

We do not fault Dawkins the biologist for a bogus handling of Shakespeare. We fault him for a bogus handling of another great work, the living organism. Ironically, his computer simulation doesn't map Darwinian evolution; it maps living things, which, like his computer program, require both an existing form or structure and a defined goal for their construction. As we intend to show, living things need to be assembled purposefully rather than one blind step at a time, for they require the right biological form or structure as the context for their genetic information to function.

To summarize, if one wanted to test the power of natural selection working on random variation—bracketing off for the moment the formal demands of the larger organism—a proper computer model would have to be truly blind, requiring whole sentences and even paragraphs to emerge from a random process in a single step without any directive goal. One might imagine that this model is more defective than Dawkins's model because such a computer program couldn't possibly assemble even a single passage from *Hamlet* or even the ghost of an imitation of *Hamlet* or any other coherent passage. But the failure isn't with the analogy but with the Darwinian mechanism. The program now maps Darwinian evolution correctly and demonstrates that Darwinian selection can't build complex, functional, organic form because, first, unlike Dawkins's program, it can't see its destination; and, second, it needs prohibitively large units of functional change to progress.

We'll examine more carefully the need for both form and goal to define biological function in the final chapters, but the argument is already at play here. Dawkins begins with a nonfunctional, meaningless jumble of letters: *WDLMNLT DTJBKWIRZREZLMQCO P.* In order to program his computer to continually approximate the target phrase, he must presuppose the target phrase itself. But the order of the target phrase itself has in reality been generated by, and has its meaning and function in terms of, the sophisticatedly integrated plot of the entire play. And so the problem is not just that Dawkins wades blithely through a whole series of dysfunctional, meaningless intermediates to get to *Methinks it is like a weasel,* but that the phrase itself is being considered in isolation from its actual existence and function as part of a whole. The error is fundamental. Dawkins presupposes what he intends to destroy; that is, he relies on the functional meaning of a sentence, crafted in relation to the larger whole of the play, to show that the purposeless, meaningless jostlings of chance can create what appears to be the act of a powerful intellect, namely, Shakespeare's. (In this, he is much like the nihilists who use meaningful words to argue for the ultimate meaninglessness of all speech and action.)

Dawkins might reply that he was only using Shakespeare to illustrate the powers of cumulative selection and that unguided nature does not have to produce entire plays; rather (to continue the literary analogy) nature begins from a couplet and works its way to a sonnet, a sonnet becomes a soliloquy, a soliloquy expands into a scene and so on until an entire play, a complex living work of art, is produced.

There are significant problems with such a rejoinder. First, Dawkins would confront the same difficulty as noted above in generating even a couplet because, to mimic cumulative selection, he must wade through a field of meaningless, functionless intermediates—from a meaningless string of original letters—and do so without bringing in the teleology of a target phrase in the computer program. Act 3 scene 3 of *Hamlet* ends with a heroic couplet spoken by Claudius: "My words fly up, my thoughts remain below: / Words without thoughts never to heaven go." If we gave the computer the run of a typical keyboard and told it to randomly generate a string of eighty-five characters including spaces, the probability of getting the precise couplet on each try would be less than one chance in 10^{150}. Even if we assume there are 100 billion, billion, billion, billion serviceable heroic couplets for the job, that still doesn't reduce the odds to better than one chance in 10^{150}.

And our plucky computer has only begun. Each more complex level would entail that it trudge through ever greater numbers of functionless, meaningless deserts in its effort to conjure up a sonnet, a soliloquy, and so on—which Darwinian natural selection wouldn't allow it to do, you'll recall, since each generation has to be functional.

When Dawkins, modern evolutionary theory's most celebrated defender, has to prop up the Darwinian mechanism with a bogus computer model, it's reasonable to suspect that something's rotten in the state of Darwin. Our suspicion is merely heightened when we choose not to bracket off the problem of organic form, a form Dawkins slips in via the intricately designed computer hardware that renders the software program functional, and by ignoring the significance that the play's larger context is essential to the meaning of the line. Moreover, the vision of the whole play was integral to the creation of the line and to each of its parts, from a couplet to an entire act. And here we don't even simply mean Shakespeare's rough sense of the story to be told as he set about to write the play. Shakespeare found his direction from a host of influences beyond the older legend on which he based his play. His own experiences, perhaps including his own father's death, the conventions of English theater, the English language of his day and the larger English Renaissance culture—all of these were essential to the generation of the work. Likewise, and as we will show in subsequent chapters, even the very simplest living cell requires a larger context that is itself astonishingly intricate. We cannot begin with a nonfunctional jumble of chemical elements and get to a functional organic unity, because natural selection cannot act unless there exists a living organic unity able to reproduce and pass on specific traits. And the simplest living organic unity surpasses *Hamlet* in terms of its integrated complexity.[12]

It is all too common, unfortunately, to believe that any living thing can be reduced to its DNA, that its DNA can then be reduced to strings of the paired nucleic "letters" A, G, C and T (adenine, guanine, cytosine, thymine) and that these letters can be reduced to their elemental constituent chemical letters, N, H, O and C (nitrogen, hydrogen, oxygen, carbon). Indeed, as we know from Dawkins's larger corpus, such a belief is at the foundation of his use of *Hamlet*. On this view, every living thing is just a string of chemical letters. But that is a fundamental error. No sensible person would say the analogical equivalent about Shakespeare's *Hamlet*. The play isn't essentially a string of letters or even a string of five acts. Its meaning is manifested from both the whole and its milieu. The error concerning DNA consists in believing that we can isolate the functional entity (the sentences or DNA) apart from that in which and by which it has its function. As with the sentence *Methinks it is like a weasel,* actual DNA is only functionally meaningful in regard to the complex unity for which it is encoded, and in and by which it can function as DNA. Our consideration of it in abstraction from its actual context is only an abstraction, a reduction useful for certain kinds of investigation but misleading and pernicious when taken literally. The cell is not, contrary to common presentation, merely a biological afterthought, a convenient container in which to keep DNA; rather, the cell is the integrated, complex whole within which DNA as DNA can function. That is, its function requires untold layers of complexity that DNA itself does not provide.

This is an important consideration when we approach any natural object and ask about its ultimate source. All too often in the Darwinian literature,

---

[12]The task that would actually face Dawkins, if one could correct his analogy, perhaps would be to move from *Hamlet,* by random variations of a few letters at a time, through a series of meaningful and excellent plays to a great postmodern novel, with no target novel in the system.

we meet with *mere pieces* of highly complex organisms, floating in isolation for inspection and judgment just like *Methinks it is like a weasel*. A floating vertebrate eye (or perhaps just the lens), floating ears, spinal columns, wings. We are then given an inviting narrative as to how such an intricate device could arise, an explanation that, since the complex piece is abstracted from the actual complex whole, *appears* to make its generation by chance at least not absurd. But this is, again, an error.

To focus on one important example, vertebrate eyes couldn't evolve in isolation precisely because eyes don't have a function as complex, self-contained entities. Eyes are part of very complex, interconnected living beings, even though they are all too often treated in the way that children treat snap-together human models, where eyes are pieces that pop in and, once having been snapped into place, allow the model to "see." But if we trace the actual bodily connections entailed in seeing, it would become clear that eyes are a very small part of the capacity for vision. Vision is the dramatic activity, the whole that defines the eye as a part.

This problem cannot be avoided by asserting that the eye can be built up gradually from a single patch of light-sensitive skin through various stages, slowly reaching the complexity of the vertebrate camera eye. Why? If you are going to make the case for the evolution of the vertebrate eye or even a light-sensitive patch of skin, the argument must be made in regard to the entire complexity of the living organism, at least insofar as that complexity supports vision (even in the least complex form). For this reason, the debate shouldn't be about the evolution of the eye, but about the evolution of vision, and vision is always the vision of some particular kind of living animal, a living whole in which the integrated activity and experience of seeing, even in its simplest form, can take place.

Such a requirement, understand, is entailed in Darwin's proper insistence that for a trait to be naturally selected, it must be functional. And it is that requirement that creates so much of the trouble for contemporary Darwinian explanations for the evolution of vision. Already the path from blindness to light-sensitive patch to vertebrate eye is hopelessly spotty, with vast oceans of unexplained development from one stage to another. But we must realize that reaching the abstract, floating world of the vertebrate eye isn't even enough. Vision is more than the eye. In vision, we are not receiving isolated patches of color, but experiencing whole objects that are colored, and we see them as existing separately (i.e., we do not see an image in our brain, but a colored thing out there). When we look out a window onto a grove of trees, we do not see varied patches of green, black and brown on which we impose order; we see *trees*. Further, it is each of us who had a desire to use his or her vision to look out the window, who chose to look up and focus on particular objects. That is the drama of vision, and it is this drama that must be the focus of the debate between Darwinists and proponents of intelligent design.

Much as we said of DNA and the cell, vision is part of an integrated, complex whole within which the eye can function. That is, its function requires untold layers of complexity that the eye itself does not provide.

Now when we say that contextual dependence is true of Shakespearean passages and then note how living structures share this quality, we don't mean Shakespeare's plays sprang whole from his mind. Obviously, when physically writing *Hamlet,* Shakespeare's pen moved, from left to right, generating one letter after another: "Methinks it is like a weasel." But the order of the play as a whole, existing in Shakespeare's intellect, influenced which letter, word, sentence and speech would follow upon which.

We are also not arguing in regard to Shakespeare that the creative process moves unidirectionally from whole to part. If the testimony of other great artists is any indication, Shakespeare's creative labor over the particular, local aspects of a given scene sharpened or brought into focus the overarching order of the play, a recursive process with the particular and the general in constant, creative tension,[13] with the whole suggesting the part and the part influencing the whole, a process simultaneously teleological and interdependent. For Shakespeare, perhaps the character of Hamlet walked into his imagination before he had begun the play, or perhaps a scrap of his opening speech. Nevertheless, a close study of the play and of the creative process among literary artists generally shows that absent a sense of the larger plot and themes—however hazy and incipient they may have been initially—Shakespeare would not have generated the particular words of Hamlet's opening speech.

That Shakespeare's art is not reducible to material mechanisms should be obvious, but the obvious has been obscured by our acceptance of philo-

---

[13]Certainly there are stories of authors seized by an opening sentence that would prove fateful. "In a hole in the ground, there lived a hobbit": it's reported that the line popped into J. R. R. Tolkien's head as he was grading a student's paper. He jotted it down, and from it grew *The Hobbit* and eventually *The Lord of the Rings*. It makes for a more dramatic telling to leave out that Tolkien had been developing a history of Middle Earth for years, all the way back to his time in the trenches of World War I.

sophical materialism. Materialism has created a kind of flatland that crushes the life out of life, despoiling its native richness, denying its true depth, mudding over its brilliant and variegated hues. In the late nineteenth-century work *Flatland: A Romance of Many Dimensions,* Edwin Abbott described a sphere that comes to visit a man in a two-dimensional world. The sphere seems to appear out of thin air on the floor, first as a point and then as a circle that waxes and wanes until it is again a point that disappears. The sphere has passed completely through the plane of the floor and now is gone. Before leaving, the sphere tries to teach the two-dimensional man the meaning of a sphere, but the Flatlander can't or won't look up, won't accept the meaning of "up." In Flatland, we learn, two-dimensional people commit suicide by drawing circles around themselves: without an up and down in Flatland, there is no escape from a circle.[14]

Fortunately, both Flatland and the materialism posing as rationalism are fiction. None of us are trapped in two dimensions, and a masterpiece like *Hamlet* is not the product of random, cumulative selection over a period of weeks or ages. Meaning isn't generated by a sequential and undirected causal chain, and neither is life.

How to escape from the flatland of materialism? We have already gathered some of the necessary tools and provisions. By understanding how much Shakespeare packs into short passages, and in understanding how these have their meaning as part of an integrated and larger whole, we acquire some of the resources necessary for understanding the inadequacy of various materialist and reductionist appropriations of Shakespeare.

These appropriations go far beyond tales of monkeys or computers randomly generating Shakespearean prose. Darwinists like Dawkins aren't the only ones who have tried to grind Shakespeare into dust. Consider, for instance, Sigmund Freud. If we see Shakespeare through the lens of a thoroughgoing Freudian criticism, suddenly we find that everything in *Hamlet,* no matter how seemingly noble or profound, can ultimately be reduced to an interplay of the two material causes of all human thought and action, the fear of death and the desire for sex. In his "Autobiographical Study," Freud wrote that one of the profound intellectual changes in his life came when he realized the ubiquity of the Oedipus complex, especially as it was man-

---

[14]Edwin A. Abbott, *Flatland* (1884; reprint, Mineola, N.Y.: Dover, 1992). John Clayton suggested the idea of *Flatland* as an analogy to spiritual blindness during a lecture in Port Orchard, Washington, in the fall of 2004.

ifested in literature. He then singled out Shakespeare's *Hamlet* as an illustration, a play which "had been admired for three hundred years without its meaning being discovered or its author's motives guessed."[15]

Contrary to the common and commonsense belief that Hamlet is naturally upset about the death of his father and the all-too-quick marriage of his mother, Freud believed that Hamlet's psychological drama was actually the result of the clash between his hidden sexual desire for his mother and his Oedipal guilt over the murder of his father. Although it was actually Hamlet's uncle who murdered his father, Hamlet's "hesitation in avenging his father," asserted Freud, is the result of "the obscure memory that he himself had meditated the same deed against his father from passion for his mother."[16] Furthermore, Hamlet, the "neurotic creation" of Shakespeare, who came "to grief, like his numberless fellows in the real world, over the Oedipus complex," was merely a dramatic shadow of Shakespeare's own inner turmoil, for "Shakespeare wrote *Hamlet* very soon after his father's death."[17]

We submit the obvious: Freud's *Hamlet* exists only in Freud's own perturbed imagination, one that pounds all within its reach into his drab, reductionist psychology. While the Freudian reading of *Hamlet* is (happily) less fashionable than it once was, the attempt to reduce Shakespeare's work to the lowest urges has not abated. In the appropriately titled *Rewriting Shakespeare: Rewriting Ourselves*—an exercise in materialism-induced relativism— Peter Erickson construes Shakespeare's *Hamlet* as a kind of adolescent assertion of masculinity in the shadow of a feminine power, namely, the queen of England. "The latent cultural fantasy in *Hamlet*," Erickson explains, "is that Queen Gertrude functions as a degraded figure of Queen Elizabeth."[18]

What then of Hamlet's tragic death, one might ask. It's not much of a fantasy. "Hamlet has to pay the tragic price of death, yet it is not only worth it but is actually pleasing and gratifying," Erickson writes. "Dying well means the ability to stage a traditional masculine self-image uncompromised by female intervention. The integrated network of gender, genre, and nation in *Hamlet* eludes Elizabeth's '*power to charm*' (1.1.168)."[19]

---

[15]Sigmund Freud, "An Autobiographical Study," in *The Freud Reader*, ed. Peter Gay (New York: W. W. Norton, 1989), pp. 38-39.

[16]Sigmund Freud, "Letter to Wilhelm Fliess, October 15, 1897," in *The Freud Reader*, ed. Peter Gay (New York: W. W. Norton, 1989), p. 116.

[17]Freud, "Autobiographical Study," pp. 38-39.

[18]Peter Erickson, *Rewriting Shakespeare: Rewriting Ourselves* (Berkeley: University of California Press, 1991), p. 86.

[19]Ibid., p. 90.

The important thing for such critics is to reduce Shakespeare. The most obvious reason is that Shakespeare was too traditional, too much the dead white Christian male, too much the theist.[20] More subtly (and this motivation need not be a conscious one in every case), Shakespeare's very art is a powerful argument against the notion that we are but the accidental outcome of a mindless material mechanism. How can you get to post-theism if to get there you have to go through such a majestic poet? It's so much easier if the Bard can be buried by a round of mud-slinging. Does somebody have a clever reading to make Shakespeare look like he's too much of an egomaniac to handle strong females—Queen Elizabeth in particular? Good! Or look over here. The Shakespeare of Leonard Tennenhouse's *Power on Display* is nasty for almost precisely the opposite failure. As Graham Bradshaw notes, in Tennenhouse the Bard of the Elizabethan era is "a sycophantic court toady whose plays were 'a vehicle for disseminating court ideology,' a man who stopped at nothing in his determination 'to idealize political authority.'"[21]

It's not difficult to play this interpretive game either. Shakespeare felt so dirty about being a "court toady" for Queen Elizabeth that he enacted a male revenge fantasy against females, calling it *Hamlet,* a subconscious reference to his hamlet of Stratford-upon-Avon where first his mother and then Anne Hathaway, the older woman who seduced him into marriage, suffocated him as a boy. Then his phallic penning of this revenge fantasy against women made him feel guiltier still, so he killed off his hero and everyone within a fifty-foot radius of the ending—Shakespeare as pathetic case of arrested development, a sort of repressed Columbine High mass murderer. In

---

[20]Some have argued that Shakespeare wasn't a Christian, pointing to portions of his work like sonnet 71, where he urges his beloved not to mourn for him "when I perhaps compounded am with clay." This is a bit like pointing to some portion of the Psalter where the psalmist struggles in doubt and confusion as evidence that the Psalter was assembled by a skeptic of monotheism. To interpret sonnet 71 in this way also reads onto Shakespeare a pre-Cartesian dualism, which regards the body as a sort of irrelevant shell for the soul, a view outside orthodox Christian doctrine on the matter. To see the Christian theism in Shakespeare, one need not look to explicit passages, such as sonnet 146 or the epitaph Shakespeare provided for his tombstone—passages where his belief in an afterlife is strongly suggested or explicitly stated. Rather, the warp and woof of his oeuvre is drenched in Christian theism. That his poetry records his struggle with faith merely places him in the line of the great authors of belief, from David through Fyodor Dostoyevsky, writers whose work captures what Blaise Pascal referred to as the grandeur and *misère* of human existence (discussed further in chap. 10).

[21]Graham Bradshaw, *Misrepresentations: Shakespeare and the Materialists* (Ithaca: Cornell University Press, 1993), p. 27. Bradshaw quotes Leonard Tennenhouse, *Power on Display* (London: Methuen, 1986).

the current anything-goes climate of literary criticism—where for some critics literature has no inherent meaning and hence can bear any meaning—
there's no need to stop there. Shakespeare was also attracted to phallic dag-
gers, one could argue. They're everywhere in *Hamlet,* and in *Macbeth* too.
And swords? Practically every one of his soldiers has one. And that name—
Shake-speare: surely a swaggering pseudonym.

Satire, yes, but is it any more ridiculous than reducing the glories of Hamlet to a young man's supposed urge to kill his father and bed his mother? A
strong impetus behind all such readings—from Freud to Erickson to Tennenhouse—is the reductionist presupposition that humans are simply material creatures and that whatever may *appear* elevated and profound to us,
whether it be Shakespeare's plays or Shakespeare himself, must ultimately
be reduced to the lowest urges.

Such readings are the natural heirs of the Darwinian account, and this
should not surprise us. Freud was deeply impressed by Darwin's theory
while later thinkers treated Darwinian evolution as a given. According to a
straightforward extension of the theory of sexual selection, which Darwin
developed to complement his theory of natural selection, the peacock has
an extravagant tail and Shakespeare an extravagant gift for spinning tales the
better to attract a mate.[22] Shakespeare's an amazing songbird, the reasoning
goes, but so what? Sexual selection has produced all sorts of extravagant
things. Just go to a zoo and see for yourself.

The philosophy of materialism is the great equalizer, and no creature escapes, for all creatures, great and small, are in such a world mere epiphenomena of the smallest matter's pointless shufflings. "Now they swarm in
huge colonies, safe inside gigantic lumbering robots, sealed off from the outside world, communicating with it by tortuous indirect routes, manipulating
it by remote control," explains Dawkins. "They are in you and in me; they
created us, body and mind; and their preservation is the ultimate rationale
for our existence. They have come a long way, those replicators. Now they
go by the name of genes, and we are their survival machines."[23]

If we imagine that the art of the Bard, or any art, is somehow exempt

---

[22]Darwin asserted that, just as the males of most other animal species have more elaborate
plumage, greater pugnacity and more developed weapons of defense, so also in human beings, males surpass females in "the intellectual powers" whether we consider the activities
"requiring deep thought, reason, or imagination, or merely the use of the senses and hands."
Charles Darwin, *The Descent of Man, and Selection in Relation to Sex* (Princeton, N.J.: Princeton University Press, 1981), 2:327.
[23]Richard Dawkins, *The Selfish Gene* (New York: Oxford University Press, 1976), pp. 19-20.

from this reductionist acid, Harvard sociobiologist Edward O. Wilson hopes to set us straight. He leads into the matter by noting that the human mind "will be more precisely explained as an epiphenomenon of the neuronal machinery of the brain. That machinery is in turn the product of genetic evolution by natural selection acting on human populations for hundreds of thousands of years in their ancient environments." A bit later he adds, "The social scientists and humanistic scholars, not omitting theologians, will eventually have to concede that scientific naturalism is destined to alter the foundation of their systematic inquiry by redefining the mental process itself." [24] And what of Shakespeare? Wilson stirs every artist into a single petri dish, proclaiming that the "sensuous hues" of art "have been produced by the genetic evolution of our nervous and sensory tissues; to treat them as other than objects of biological inquiry is simply to aim too low." [25]

Wilson puts a brave face on his recommended approach, implying as he does that his opponents are aiming "too low," but his attempts to apply his reductionist explanation to the riches of Shakespeare lead not into Shakespeare but into the materialist wasteland, the stultifying flatland of modernity at its most clinical, lifeless and, hence, most nihilistic.

This reductionist urge in both Shakespearean studies and biology, this fixation on hypertidiness, no doubt has a variety of historical antecedents. One of special significance is the metaphor of the cosmos as watch (one crafted, ironically, to underscore how elegantly designed the universe is).

We can see how Enlightenment thinkers arrived at this metaphor, confronted as they were with fresh insights concerning the mathematical regularities in the laws of motion and among the relationships of celestial bodies. And today in astrophysics, even those who resist the idea of cosmic design now tell us that the laws and constants of the cosmos are finely tuned to an almost unimaginable degree, such that even small changes to them would render any sort of complex life impossible. So at least in one important sense, the universe is watchlike: its physical constants resemble a precision instrument.

The impact that the watch metaphor had on the arts isn't difficult to trace. The influences leading to the period in the arts we call neoclassical are, of course, complex. But one obvious influence moving artists from the sprawl-

---

[24]Edward O. Wilson, *On Human Nature* (Cambridge, Mass.: Harvard University Press, 1978), pp. 195, 204.
[25]Ibid., p. 11.

ing richness of Renaissance drama to the tidy couplets of a John Dryden or an Alexander Pope was precisely this insight into the mathematical and seemingly watchlike regularity of the cosmos. The trouble comes when the analogy is reified. It is an illuminating analogy but only an analogy. Where a single metaphor crowds out all others in a matter as complex as our living world, it produces an intellectually impoverished, misleading stick figure of the subject—an abstract, straw theory that can produce abstract, straw men. Thus, the thinking person is wise to ask, to what extent is the universe watchlike? To what extent should it be watchlike? All metaphors break down if pressed far enough, and this one breaks down pretty quickly.

As heirs of the Enlightenment, our imaginations have been improperly constricted by this mechanistic analogy. What do we find elsewhere, further back? Whether we think of the morally compromised gods of Mount Olympus meddling in the affairs of their various mortal offspring; or of Plato's "the One" (what he also called "the Good" or "Father of that Captain and Cause"); or of the holy God of the Bible, father and shepherd and husband of his people, the deity is not construed as one interested in the world primarily as a tool for himself, much less as a watch to display merely mechanistic regularity. Indeed, whenever he is construed as a personality, and not merely as some sort of nonsentient organizing First Principle, he is depicted as one interested in the world itself, as a creator who delights in the work of his hands.

Dare we use the word *love* in this context? Dare one suggest that the designer might love his creation, that he might no more think of his creation as a tool or machine than would a father his child, or Shakespeare his art? That such a term as *love* strikes many as inappropriate in an origins debate merely testifies to how thoroughly the utilitarian and reductionist implications of the watch metaphor have permeated Western thinking.

Biologist and leading Darwin defender Kenneth Miller comments,

> The God of the intelligent-design movement is way too small. . . . In their view, he designed everything in the world and yet he repeatedly intervenes and violates the laws of his own creation. Their God is like a kid who is not a very good mechanic and has to keep lifting the hood and tinkering with the engine.[26]

Miller isn't a materialist, but in this respect he takes the materialists' view

---

[26]Kenneth Miller, quoted in Paul Nussbaum, "Evangelicals Divided over Evolution," *Philadelphia Inquirer,* May 30, 2005, p. A1.

of how God, if there were a God, should behave. Notice how blithely he equates the designer's ongoing involvement in creation with incompetence. But what if the creator likes to stay involved? What if he doesn't want to wind up the watch of the cosmos and simply leave it, disinterestedly, to wind out every stage of creation from supernovas to sunflowers? What if he wants to get his hands dirty making mud daubers?

What if, as church father Gregory Nazianzen suggested in the fourth century, we compare the universe to a lute played by the lute's maker?[27] Or what if the designer is more like a wild dramatist than a tame watchmaker? Would we say to the Bard, "You keep writing and rewriting your plays! You have the unhappy imperfection of wanting to stage your creations with live actors, and even worse, of directing them! You repeatedly violate the laws of drama and poetry with your detestable and irrepressible urge to create something new—absolutely unable to leave well enough alone! You stoop too low for the sake of the groundlings and reach too high for the sake of the philosophers! You are the shame of the theater!" Such are many of the criticisms of the designer by those seeking to dismiss the very idea of one. And no wonder. The living world is more like Shakespeare's Globe Theater than the tidy and tightly bound circle of the watch.

Certainly, we could try to discuss the order of nature without considering the designer's attitude toward his work (that is, whether he is more a watchmaker, bridegroom or dramatist). But the Darwinists have already smuggled this issue into the debate by assuming that, if there were a designer, he could only be some sort of disinterested and hypertidy engineer. Having smuggled it in, they then regard as beneath consideration any evidence of a designer who (as they put it) "meddles in his creation."

Similarly, they dismiss the notion that an omnipotent and omniscient designer might fashion a creature that, considered narrowly, would seem to fall short of an optimal design. Here they not only make a theological claim

---

[27]Gregory Nazianzen *Orations* 27-28, in *Nicene and Post Nicene Fathers of The Christian Church,* 2nd ser., vol. 17, ed. Philip Schaff and Henry Wace (Peabody, Mass.: Hendrickson, 1994). Available at Catholic First: Catholic Information Center on the Web <http://www.catholicfirst.com/thefaith/churchfathers/volume30/gregory3005.cfm>. The relevant passage reads, "For how could this Universe have come into being or been put together, unless God had called it into existence, and held it together? For every one who sees a beautifully made lute, and considers the skill with which it has been fitted together and arranged, or who hears its melody, would think of none but the lutemaker, or the luteplayer, and would recur to him in mind, though he might not know him by sight. And thus to us also is manifested That which made and moves and preserves all created things, even though He be not comprehended by the mind."

but ignore key questions at once practical and aesthetic: How do concerns about ecological balance impinge on a critique of animal structures? Or more poetically, how does each creature play a part in the overall drama of life? Again, as with Dawkins and his sentence-generating computer program, they overlook the larger context. They fault the designer, for instance, for not giving pandas opposable thumbs. An omniscient and omnipotent designer would already have known about the superior opposable thumb, they argue, and would have been sure to give it to them. Since he didn't, he obviously doesn't exist or, at least, isn't directly involved.

But must the cosmic designer's primary concern for pandas be that they are the most dexterous bears divinely imaginable? The sturdy thumbs they do have work beautifully for the work set before them, peeling bamboo. From a purely practical standpoint, might not opposable-thumbed über-pandas wreak havoc on their ecosystem? And from a purely aesthetic standpoint, might not those charming pandas up in the bamboo with their unopposing but quite workable thumbs be just the sort of humorous supporting character this great cosmic drama needs to lighten things up a bit? If Shakespeare could do this sort of thing, why not the maker of Earth?[28]

Pandas as comic relief, as divine whimsy? To spurn the notion as if it were patently ridiculous, as beneath consideration, is merely to expose one's utilitarian presuppositions. Why, after all, should the designer's world read like a dreary high school science textbook, its style humorless, homogenous and suffocating under the dead weight of a supposedly detached passive voice? Why should not the designer's world entertain, amuse and fascinate as well as "work"? Why, in short, shouldn't we expect it to have the richness of variety and tone we find in a work of art like *Hamlet*?

The reason many do not is that the bad-design versus good-design discussion is often framed by an engineer's perspective, not an artist's or mystic's. When I (Jonathan) noted this to philosopher Jay Richards a few years ago, he responded in a letter:

After all, why do we assume that God created the universe to be a watch, in which a self-winding mechanism makes it "better"? Maybe the universe is like

---

[28]The material about the watch metaphor and the defective aesthetic arguments against design is drawn from Jonathan Witt, "The Gods Must Be Tidy! Is the Cosmos a Work of Poor Engineering or the Gift of an Artistic Designer?" *Touchstone*, July/August 2004, pp. 25-30, material the author previously presented at a December 2003 conference of fellows of Discovery Institute's Center for Science and Culture and again at an April 2004 conference at Biola University.

a piano, or a novel with the author as a character, or a garden for other beings with whom God wants to interact. It's amazing how a simple image can high-jack a discussion for a century and a half.[29]

The reductionist-watchmaker thinking of evolutionists like Stephen J. Gould and Dawkins paves the way to all sorts of unwarranted conclusions. Gould preaches against this atomistic view that "wholes should be under-stood by decomposition into 'basic' units" but nevertheless holds it himself. He and many other biologists assume not only that nature is a kind of watch, but that each individual design is its own watch—its own machine—meant to be judged in relative isolation. They evaluate the panda's thumb by how well it works as a thumb, not by how well it fits into the whole life of the panda, including its place in its own environment. This is, at the most prac-tical level, to misunderstand pandas. At the aesthetic level, it is to declare that an artist who might have created pandas could not have been thinking (as artists do) of the whole work. It's the same mistake they make when they treat DNA in isolation from the cell, and the eye in isolation from vision. This is not an idle charge. Dawkins ignores the larger demands of vision in his critique of the vertebrate eye, zeroing in on the eye's so-called backward wiring:

> Each photocell is, in effect, wired in backwards, with its wire sticking out on the side nearest the light. . . . This means that the light, instead of being granted an unrestricted passage to the photocells, has to pass through a forest of connecting wires, presumably suffering at least some attenuation and dis-tortion (actually probably not much but, still, it is the *principle* of the thing that would offend any tidy-minded engineer!).[30]

His analysis collapses under two mistakes. First, it has been clearly dem-onstrated that the backward wiring of the mammalian eye actually confers a distinct advantage by dramatically increasing the flow of oxygen to the eye.[31] Dawkins the reductionist misses this because he tends to analyze or-gans and organisms in the same way that he treats phrases in Shakespeare—in isolation (at least when it suits his purposes). Then there is Dawkins's ob-session with neatness. O brave new world whose supreme designer distin-

---

[29]Jay Richards, e-mail to Jonathan Witt, fall 1999.
[30]Dawkins, *Blind Watchmaker,* p. 93.
[31]See Michael J. Denton, "The Inverted Retina: Maladaptation or Pre-adaptation?" Origins and Design 19, no. 2 (winter 1999): 14. This is available online at Access Research Network <http://www.arn.org/docs/odesign/od192/invertedretina192.htm>.

guishes himself principally by a reductionism masquerading as tidyness! Novelist Aldous Huxley ably dramatized the horror of a society so engineered. Do we really wish to substitute Dawkins's "tidy-minded engineer" for the exuberantly imaginative, even whimsical designer of our actual universe? Such a deity might serve nicely as the national god of the Nazis, matching Adolph Hitler stroke for stroke—Hitler in his disdain for humanity's sprawling diversity, the tidy cosmic engineer in his distaste for an ecosystem choked and sullied by a grotesque menagerie of strange and supposedly substandard organs and organisms. Out with that great big prodigal Gothic cathedral we call the world; in with a modern and minimalist blueprint for a new and neater cosmos, a drab but efficient flatland of utility.

The god of the English canon, Shakespeare, has received much the same criticism from the tidier neoclassical critics as the Author of the cosmos has received from the hypertidy scientists of our present age. This actor turned-playwright lacked classical restraint, the argument went. Lewis Theobald perhaps initiated the century's long criticism of Hamlet's coarse speech in 1726 when he commented on a particularly bawdy line spoken by Hamlet to Ophelia: "If ever the Poet deserved Whipping for low and indecent Ribaldry, it was for this Passage."[32] Another critic regarded Shakespeare's general habit of mingling the low with the high, the comic with the tragic, as a "wholly monstrous, unnatural mixture."[33] With only a little more restraint, a third lamented the Bard's tragedies: "How inattentive to propriety and order, how deficient in grouping, how fond of exposing disgusting as well as beautiful figures," how often he compels the audience "to grovel in dirt and ordure."[34]

As one modern critic noted, even the admiration of the more sympathetic neoclassical critics was always "modified and tempered . . . by regrets that Shakespeare had elected, either through ignorance or by design, to embrace a method that discarded all classical rules."[35]

---

[32]Lewis Theobald, quoted in Paul S. Conklin, *A History of Hamlet Criticism: 1601-1821* (New York: Humanities Press, 1968), p. 53.

[33]Charles Gildon, quoted in Herbert Spencer Robinson, *English Shakespearian Criticism in the Eighteenth Century* (New York: Gordian Press, 1968), pp. 26–27.

[34]Edward Taylor, "Cursory Remarks on Tragedy, on Shakespeare, and on Certain French and Italian Poets, Principally Tragedians" in *Shakespeare: The Critical Heritage, 1774–1801*, ed. Brian Vickers (London: Routledge & Kegan Paul, 1981), pp. 130–32. "Cursory Remarks" was published anonymously and is most often attributed to Edward Taylor. This Taylor is not to be confused with the brilliant American poet Edward Taylor, the last of the metaphysical poets, who spent a great deal of time in the "dirt and ordure" exploring the mysteries of the human and divine.

[35]Robinson, *English Shakespearian Criticism*, p. xii.

What do we make of such criticism today? Most find it damagingly narrow. What emotionally whole and thoroughly sane admirer of Renaissance drama would want to replace the works of the "myriad minded" Shakespeare with the relatively impoverished fare left over after unsympathetic neoclassical critics tidied him up? We can of course say the same thing about the Shakespeare that Freud leaves us, or the one Dawkins or Erickson or Tennenhouse suggests. They've tidied him up by crushing him down into the neat little boxes of materialism, and there's very little blood or humanity left for the detective who follows after: the universal acid of Darwinism cleans everything up spic and span.

The full relevance of our comparison should now be clear. The reductionist treatments of Shakespeare noted above are akin to the Darwinists' overly tidy treatment of vision or the cell. In each case the critic analyzes the work narrowly, ignoring the larger context, be it ecological, aesthetic or otherwise. Proponents of this line of argument value a hyperconstricted and abstract elegance over other and often more vital attributes like variety, imaginative exuberance, freedom and even moral complexity. In their attempt to master everything, they deny anything that exceeds their grasp. They lose the meaningful whole. If that's lucidity, then it is a kind of mad lucidity.

A century ago, G. K. Chesterton described the materialist in very much these terms, one who has become less than human by binding everything by the chains of his own cramped circle of logic:

> Now, speaking quite externally and empirically, we may say that the strongest and most unmistakable *mark* of madness is this combination between a logical completeness and a spiritual contraction. The lunatic's theory explains a large number of things, but it does not explain them in a large way.[36]

So also with the materialist. "As an explanation of the world, materialism has a sort of insane simplicity," continues Chesterton. "It has the quality of the madman's argument; we have at once the sense of it covering everything and the sense of it leaving everything out."[37] The error of the materialist, we're arguing, is a type of delusion, the constriction of reason within a tight little circle. We say with Chesterton to those who have fallen into materialism, "I admit that your explanation explains a great deal; but what a great deal it leaves out!"[38]

---

[36]G. K. Chesterton, *Orthodoxy* (Garden City, N.Y.: Image, 1959), p. 20.
[37]Ibid., p. 23.
[38]Ibid., p. 20.

The reductionist thinking Chesterton takes aim at is easier to avoid where one allows for the possibility not only that the universe is designed, but also that the creative imagination behind that design—like Shakespeare's imagination—is not less but more than the average human's, more than the average scientist's. Perhaps the excellence of creatures like the panda or the duck-billed platypus is that they are even more fanciful—and certainly articulated to a far greater degree—than the creatures of our imagination. As Chesterton also says, it is one thing to consider "a gorgon or a griffin, a creature who does not exist. It is another thing to discover that the rhinoceros does exist and then take pleasure in the fact that he looks as if he didn't."[39]

The materialist bent on reducing everything scoffs at talk of a whimsical designer, insists on viewing the splendor of both Shakespeare and the world with a kind of stubborn morbidity. What to do? "If you or I were dealing with a mind that was growing morbid," Chesterton suggests, "we should be chiefly concerned not so much to give it arguments as to give it air, to convince it that there was something cleaner and cooler outside the suffocation of a single argument."[40]

Nor is the dogmatic atheist the only one in need of such medicine. Materialism is now part of the cultural air we breathe. In the next chapter we escape its fumes by taking the sea air, traveling to the island of Shakespeare's *Tempest* and exploring how his art calls us to appreciate a range of aesthetic qualities far more various than the single argument of a tidiness wound into itself. These qualities are the elements of genius and, as we will argue, they are found not only in Shakespeare, but in nature as well.

---

[39]Ibid., p. 11.
[40]Ibid., p. 20.

3

SHAKESPEARE AND
THE ELEMENTS OF GENIUS

Oh wonder! . . . Oh brave new world

That has such creatures in't!

The Tempest

AS SHOULD BE EVIDENT FROM THE LAST CHAPTER, many of today's leading Shakespeare scholars reject theism. The awkward thing for them is that William Shakespeare's work—the work they have dedicated their professional lives to—does not. The playwright's themes pose a profound challenge for materialism, assuming as they do the ontological categories of flesh and spirit, good and evil, heaven and hell. But more fundamental still is the challenge Shakespeare's genius poses to any worldview that would reduce everything, including the human mind, to the mindless flux of matter and energy. Not only does his genius seem irreducible to anything so mean, the fruits of that genius find a striking correspondence in the ingenious forms of nature.

"Shakespeare is the Spinozistic deity—an omnipresent creativeness," exclaimed the poet Samuel Taylor Coleridge.[1] The dramatist Ben Jonson was equally effusive, exclaiming in verse that "neither man, nor muse" could praise Shakespeare too much, and that "he was not of an age, but for all time."[2] When in 1933 the editor of a Shakespeare collection commented that "every form of genius . . . may be found in the genius of Shakespeare, concentrated and con-

---

[1] Samuel Taylor Coleridge, from "Table Talk," *Coleridge's Writings on Shakespeare,* ed. Terence Hawkes (New York: G. P. Putnam's Sons, 1959), p. 92.

[2] Ben Jonson, "To the Memory of My Beloved, the Author, Mr. William Shakespeare, and What He Hath Left Us," in *The Complete Works of William Shakespeare,* ed. Farquson Johnson (New York: World Syndicate, 1933), p. lv.

densed,"[3] he was speaking a truism, the settled view of a playwright who, for more than three hundred years, had held the world in awe.

Are we really to believe that natural selection moved from a single cell in a dirty pond to this? Vague just-so stories about nature selecting for verbal skill are one thing, but standing in steady contemplation before one of Shakespeare's mature literary works and then entertaining such an explanation without feeling a deep sense of skepticism—that's a very different matter.

As we saw previously, some materialists have coped by attacking Shakespeare's reputation. Like the sharks in *Moby Dick,* consuming themselves in their bloodlust over a dying whale, some literary critics, in lockstep with the so-called death of god movement, have turned on and devoured their own field of literary studies. There is no "literature," postmodernists like Robert Scholes argue. Or rather, the category of "literature" is a fiction designed to "smother" students under a traditional cultural authority.[4] According to such a view, there are only texts, some carrying the name "literature" the better to manipulate you.

By deconstructing the binary of literature/nonliterature, the cultural critic is free to bring into a "literature" class everything from *Hamlet* to hip-hop: "Phoenix College Students Choose Shakespeare or Tupac Shakur," the institution's press release proudly proclaims; it then goes on to lay out a program right out of the pages of Scholes's *Textual Power,* where students are encouraged to define their own interests—a literary flatland where everybody is a genius, so no one is.[5] One could write off Phoenix College as a pseudo-college desperate to get students in the door, but the trend, if less proudly advertised, runs throughout academia, involving some of our most prestigious universities. And even where it's fed merely by a desire for dollars, the effects of materialism are no less at work. Universities are learning to be "customer oriented," to feed the consumer what the consumer wants, the better to allow the predatory university to feed off of the consumers feeding on them. The deconstructionist school of Jacques Derrida is much given to protesting mass-market capitalism and its creep into the academy, but these materialists have no philosophical ground to stand on because they are standing on Darwinian

---

[3]Farquson Johnson, "The Growing and Perpetual Influence of Shakespeare," in *The Complete Works of William Shakespeare,* ed. Farquson Johnson (New York: World Syndicate, 1933), p. viii.

[4]Robert Scholes *Textual Power* (New Haven: Yale University Press, 1986), chap. 1.

[5]Phoenix College, press release dated January 13, 2004 <http://www.pc.maricopa.edu/news/January%2004/literature.html>.

ground. Darwinian materialism leads rather naturally to the materialism of the mall—life as appetite. The Sprite commercial sums up the prime directive— "Obey your thirst"—and in a world where matter is all that matters, why not?

It's enough to make one empathize with the melancholic Hamlet of act 1: "How weary, stale and flat" is a world in which there is neither better nor worse, higher or lower, noble or base—where texts can mean anything precisely because they mean nothing.

We have fallen down the rabbit hole of unmeaning and had better take our bearings. Many springs and rivulets feed the great river of materialism/relativism/nihilism, but since the Victorian era, its principal tributaries are Freudianism, Marxism and, above these, Darwinism. As we discussed, if one presupposes that Sigmund Freud was correct, then everything in Shakespeare must be reduced to two motives, the fear of death and the desire for sex, no matter how elevated and removed from either it may appear. If we presuppose that Karl Marx was correct, then everything in Shakespeare can be reduced to the pushings and shovings of human beings variously related to the material modes of production, caught and defined in every thought and deed by the shaping force of class struggle.

Finally, if we presuppose that Charles Darwin was correct, then everything in Shakespeare must be reduced further still to the desire to survive and propagate, caused proximately by Shakespeare's own desire to use his lucky genetic variations to attract the ladies with his flair for poetry (a more elaborate rendition of a male bird's seductive mating call) and caused more distantly by some remote genetic ancestor, a primitive troubadour whose musical ululations, gibbered around a dying fire, skillfully wooed his audience of spellbound females. The most ruthlessly consistent and unsparing materialist will dissolve Shakespeare and his fellow geniuses to the pointless concatenations, writhings and bursts of matter and energy. In such a world, the genius is nothing more nor less than what the poet T. S. Eliot described in "The Hollow Men":

> The stuffed men
> Leaning together
> Headpiece filled with straw. . . .
> [Their] dried voices . . . quiet and meaningless
> As wind in dry grass
> Or rats' feet over broken glass.[6]

---

[6]T. S. Eliot, "The Hollow Men," in *T. S. Eliot: Collected Poems 1909-1962* (New York: Harcourt Brace Jovanovich, 1963), pp. 79-82.

This is a rabbit hole without whimsy or light.

Fortunately, in the wake of all such attacks, Shakespeare's genius remains, and his art can help to lead us out of the darkness of nihilism if we will but contemplate it. Unfortunately, at this point our efforts are apt to come under friendly fire. Many nonacademics philosophically opposed to materialism believe they are striking a blow for common sense when they dismiss Shakespeare as a flowery and irrelevant relic, insisting that we had better stick to hard science instead. But it is precisely from within the cave of materialism that Shakespeare's genius appears flat, colorless and remote from our own concerns.

Granted, contemporary readers face a real language barrier. Reintroducing ourselves to Shakespeare takes work. The principal meanings and emotions of his plays were readily assimilated by uneducated groundlings four hundred years ago, but today every eighth or ninth word is strange. A contemporary reader can overcome this, but only if she believes Shakespeare's gifts are worth the trek across the desert of initiation. Unfortunately, the modern outlook that Darwinian materialism helped to shape is no help here. It teaches us to expect nothing in the Bard worth the effort of close reading; it teaches us that Shakespeare is old and that newer is cleaner is better. Just look at the absurd flexibility of Shakespeare's syntax! What an earlier age saw as the supple freedom of genius, the modern poet came to regard as cheating, with Ezra Pound going so far as to outlaw inversions of common word order in his list of "Don'ts."[7] And in those tragedies, all of the characters take themselves so seriously. Think of Ophelia, whose father is inadvertently slain by her lover Hamlet. She goes insane, mooning about the castle day after day, talking nonsense. As for the Bard's difficult comedies, isn't laughter just letting off a bit of steam (as Freud thought), an activity that has no business demanding effort? And those required courses on Shakespeare—why can't colleges focus on practical subjects, things that'll help

---

[7] See Ezra Pound's letter to Harriet Monroe, January 31, 1915. Also consider his approval of William Butler Yeats for a poetic idiom "without inversions" in "A Retrospect," in *Literary Essays of Ezra Pound*, ed. T. S. Eliot (New York: New Directions, 1968), p. 12. Notice that in the same breath he compliments Yeats for making "our poetic idiom a thing pliable." How can an idiom be pliable that has grown too brittle to allow for inversions? In truth, Yeats stopped fighting the growing brittleness and worked skillfully within its narrowing parameters whereas many a lesser Victorian poet borrowed from the older, inversion-rich idiom just enough to sound stilted. Gerard Manley Hopkins was the notable exception. He did not borrow. He followed William Faulkner's advice and stole, and not timidly. He pillaged all the way back to the realm of the Saxons and then made those methods his own.

students get ahead when they're out in the real world?

The real world? Shakespeare is all about the real world. What the utilitarian student has in view is a kind of two-dimensional flatland, a modern culture leveled by materialism. Students with this dismissive view of English courses are voicing a materialist outlook whether they realize it or not. What if, instead of being passively carried along by such presuppositions, we actively cast them off and go to Shakespeare hoping to find what generations past expected to find—"God's plenty."[8]

Such an approach means casting off the hip disdain now all but ubiquitous in our pop culture. To this end, literary theorist George Steiner has attempted to revive two decayed words, "civility (a charged word whose former strength has largely left us) towards the inward savour of things" and cortesia, the root for the impoverished term courtesy.[9] We need cortesia and civility, he argues, in the rich, old sense of these words. Recall how lavishly the patriarch Abraham welcomed the three strangers, killing for them the fatted calf. Steiner is seeking to revive this old custom of elaborate hospitality and urging us to treat the works of the great authors as our invited guest.

But as the hospitable know, the guest, being a stranger, may prove dangerous. Here the danger is to break the spell of materialism, a danger we believe is all to the good. To that end we continue with Shakespeare, turning now to a romance chock full of spells laid and spells broken, The Tempest. Begin, if possible, by both reading and watching it.[10] Shakespeare's plays are, first and foremost, dramatic spectacles, full of vivid costumes, exciting actions, artful gestures and clever theatrical devices that enhance the viewer's understanding of character and action. The genius of Shakespeare is not abstract and aloof, removed from all but the most learned. Rather he is accessible on the common level, touching the beliefs, delights, fears and desires of an everyday audience, even while elevating them. And when we

---

[8]John Dryden used the term to describe Geoffrey Chaucer's Canterbury Tales in Dryden's preface to The Fables (1700).

[9]George Steiner, Real Presences (Chicago: University of Chicago Press, 1989), pp. 147-48.

[10]A good live performance is best, but even watching a clunky video production can help enormously, Derek Jarman's 1979 version takes such liberties with the original that it may do more to confuse than illuminate. The version titled Prospero's Books also has little compunction about shifting all manner of things around, changing speakers and generally removing actors' clothing for cheap shock value. I (Jonathan) found a 1980 production by John Gorrie preferable to another relatively faithful version I watched some years before. The production values are modest and Gorrie felt compelled to present his spirit characters as a kind of gay fraternity; all the same, the work illuminates much that might remain obscure from a textual reading alone.

see the plays, not only are we digesting them in the form Shakespeare intended, but our task of interpreting Elizabethan English is helped along by the gestures, actions, costumes and props.

The temptation at this stage of the reading is to think, "Just give me the point." But our contemporary rage for "the point" is reductionism in a business suit, materialism's urge to boil everything down to its simplest parts. This can be a constructive mode of inquiry, but where it becomes an idol, it can transform a vibrant culture into something flat and dreary. At the altar of reductionism we tend to accept grand and abstract conclusions about something without having first attended to the thing in question. It is easy to believe the reduction of butterflies to atoms and energy if we have never attended to a butterfly in the theater of the fields. Likewise, it's possible to believe reductionist treatments of Shakespeare if all one has read is the reductionist treatments themselves, avoiding the very theater within which we might be delivered from such misreading. In short, we need to *experience* Shakespeare's genius. Having done so, we will be immune to the materialist's reductive treatments of the Bard and, to some degree, immunized against the violently reductive treatments of nature and ourselves as well.

The Tempest is set on a remote island in the Mediterranean. Through the scheming of his brother (Antonio) and the king of Naples (Alonso), the true Duke of Milan (Prospero) has been banished there with his young daughter (Miranda). The only denizens of the island are several spirits, the chief of whom is Ariel; and the monstrous Caliban, the offspring of a witch. Prospero bides his time, bringing up his daughter while he awaits a providential chance for return. As the play opens, a tempest shipwrecks the very malefactors responsible for his exile on the island, along with King Alonso's son, Prince Ferdinand, and several others. In the course of the play, Prospero uses magic to bring his enemies to repentance and unite his house with that of the King of Naples through the marriage of Miranda and Ferdinand. Simple enough, but what a world of wonder Shakespeare packs into this simple frame.

It begins and ends in the romance of the word, in the ingenious beauty of Shakespeare's speech. Imagine, for example, if Miranda's opening speech in the second scene had read thus: "If you caused this storm, father, stop it! You are causing a flood, and I am concerned that that ship will crash and the innocent men on it will die. If I had the power, I would go to considerable lengths to save them." Instead we are thankful to read:

If by your art, my dearest father, you have
Put the wild waters in this roar, allay them.
The sky, it seems, would pour down stinking pitch,
But that the sea, mounting to th' welkin's cheek,
Dashes the fire out. O, I have suffered
With those that I saw suffer: A brave vessel,
(Who had no doubt some noble creature in her)
Dashed all to pieces. O, the cry did knock
Against my very heart! Poor souls, they perished.
Had I been any god of power, I would
Have sunk the sea within the earth or ere
It should the good ship so have swallowed and
The fraughting souls within her.[11]

The beauty of this passage shines out even on a first reading. It is the work of an artist at the height of his powers, and in the analysis of his style that follows, there exists a striking parallel between mistakes made in describing Shakespeare's creative process and the mistakes some scientists make in trying to explain the language of life.

Shakespeare was long depicted as an untutored and spontaneous genius: poetry gushed from him, waiting for neither effort or training. The notion was part of the Shakespeare of legend, now largely debunked by historians. Notice, however, that it's the same romantic notion that inspired a belief in undirected evolution in the decades leading up to *Origin of Species*. Long before Darwin offered a credible mechanism, there were those who argued that advanced life forms had simply and spontaneously unfolded from simpler ones, which in turn had emerged from nonlife, all without the input of a designing intelligence.

Such an outlook appears so free and easy on the surface, but notice that it puts a fresh demand on both Earth and artist. The first must create without a creator while the latter must cover the tracks of his effort lest his reputation be diminished. The very brand of romanticism that inspired evolutionary speculations in the period leading up to Darwin has driven many an artist to paper over the sweat and calculation—the intelligent design—that goes into his art. Some of the leading English poets from the Romantic period, for

---

[11]Shakespeare, *The Tempest: A Norton Critical Edition* (New York: W. W. Norton, 2004), 1.2.1-13. The phrase "th' welkin's cheek" (line 4) could be paraphrased as *the face of the sky;* 'the fire" (line 5) as *lightning;* "or ere" (line 11) is *before;* "The fraughting souls within her" (line 13) are the people on the ship construed as the ship's freight.

instance, boasted of poems coming to them fully formed, as if the sweat of design were somehow beneath them. Later scholars found rough drafts from some of these poems.

Similarly, Louis Armstrong, famous for improvising brilliant jazz recordings on the spot, actually worked out his pieces in meticulous detail beforehand, practicing them till they sounded absolutely natural and only then recording them.[12] Artists, of course, have been "struck by lightning," producing excellent works almost as if by dictation. But even here, inspiration is only half the story. In *The Courage to Create,* Rollo May shows that the famous examples of artistic or scientific inspiration were always pre ceded by long stretches of intense mental effort on the problem—whether it was Einstein puzzling over the problems related to relativity theory or a poet sweating over her craft.[13]

A related misconception that some origin-of-life researchers and students of Shakespeare share is the notion that some tidy little formula can explain great art. We will investigate the biological side of the coin in a later chapter. Here, we will develop the argument through a closer look at Shakespeare's poetic style.

The Shakespearean passage quoted above, like most of Shakespeare's dramatic work, is written in blank verse, that is, unrhymed iambic pentameter. Iambic refers to an accentual rhythm of one unstressed syllable followed by one stressed syllable. Pentameter means there are five unstressed/ stressed syllable pairs per line. Simple enough—no rhyming, just alternate stress and unstress, sort of like a metronome. The problem, of course, is that metronomes are boring, and Shakespeare didn't employ iambic pentameter to bore people. The successful use of iambic pentameter demands enormous expertise and input from the mind of the poet, much as certain mathematically or geometrically elegant patterns in the living world alone do not account for living things, but also demand a noncompressible, aperiodic kind of form as well. In both cases, the pattern, expressible in mathematically concise terms, demands an infusion of noncompressible genius before the organic whole, of which the pattern is but a part, can spring to life.

---

[12]This according to an NPR interview with a leading jazz historian, the reference for which escapes me (Jonathan). The essential point, however, is found elsewhere: "The first step away from this trap of polarities is to recognize that black music is not the universal unconscious or the primitive body projected by romanticists of various stripes but rather a highly disciplined set of practices." Susan McClary and Robert Walser, "Theorizing the Body in African-American Music," *Black Music Research Journal* 14, no. 4 (1994): 75-84.

[13]Rollo May, *The Courage to Create* (New York: Bantam, 1975).

We can imagine, then, that a novice told to write a poem in iambic pentameter might produce something like the following (the stressed syllables are placed in small caps for ease of scanning):

My DOG is BIG and BOLD and BORED to ME.
He LIES aBOUT the HOUSE and YAWNS and SLEEPS.
I LOVE him SO. I LOVE his TOES, you KNOW.
A DOG'S a FRIEND in NEED, a FRIEND inDEED.

Microsoft's thesaurus incorrectly lists *blank verse* as a synonym for *doggerel*. The thesaurus's authors might have had this sort of blank verse in mind when they conflated the two terms. Our "Ode to a Dog" is awful poetry—trite, monotonous, vague. In the third line we even manage an off rhyme that's also a forced rhyme, and this in a verse form that doesn't even demand rhyme. The offending word? *Toes*. Dogs have claws, paws, even paw pads, but not *toes*. The word was slipped in merely because it rhymed, never mind that it didn't belong. And the dog in question could be almost any largish, seemingly bored dog on the planet. There's virtually no specificity, no evidence we know our dog from Adam's. We even seem to run out of things to tell you about our dog half way through the quatrain, so much so that we have to resort to cliché. Do we even own a dog? Has Benjamin Wiker or Jonathan Witt ever so much as kept a friend's dog for the weekend? One would be hard-pressed to tell from this poem.

Notice, too, what might appear (if narrowly considered) a strength of the poem. We were supposed to write a quatrain of blank verse and we have done so, *flawlessly*. Every stress is in the right place; every unstress is properly flanked by its assertive opposite. Compare our poem's precision to the following disorderly sample (syllables in violation of pure iambic pentameter are in italics):

IF by your ART, my DEARest FAther, YOU *have*
PUT the *WILD* WAters IN this ROAR, alLAY *them.*
The SKY, it SEEMS, would POUR *DOWN* STINKing PITCH,
BUT that the SEA, *MOUNT*ing *TO* th' WELKin's CHEEK . . .

Shakespeare botches strict iambic rhythm in every line. The only thing he has going for him is that his verse is rich, textured, beautiful, vivid and, well, rhythmic. It's rhythmic not in spite of the "violations" to strict iambic pentameter, but partly because of them. Shakespeare's pen understood something our doggy poem misses: iambic pentameter isn't a mathematical algorithm somebody can just plug into and follow on autopilot. Rather, it's an underlying theme on which the master poet works subtle variations— neither so many variations that the underlying pattern is lost, nor so rarely that the poem begins to sound like a hypnotizing metronome. Instead, Shakespeare works his variations just often enough and with such skill that he lends a musicality to his language while drawing from the listener's expectation of pattern a range of subtle effects.

Take one example. In the fourth line, the underlying pattern calls for an unstressed syllable after the word *sea*. Instead there comes "MOUNTing TO th' WELKIN's CHEEK." Just as the content of Miranda's words speak of the sea mounting up, so too the line mounts up, stressed syllable piled onto stressed syllable.

Notice, too, Shakespeare's use of alliteration in the third line, that is, the repetition of words with the same beginning consonant sound. The three words are *sky, seems* and *stinking*. Now *stinking* isn't an especially pretty word, but it's the right word in the right place: Miranda isn't describing a lovely rain shower but a brutal, ugly storm. Here, however, we want to pay special attention to how the alliteration itself differs from what we find in the doggy doggerel. The first line of doggerel alliterates the words *big, bold* and *bored*. This line is monotonous, and not just because of the unimaginative repetition of the word *and* or the emerging metronomelike precision of the iambic pattern. Notice that each word starting with *b* begins with a solitary consonant followed by a vowel. Shakespeare, by contrast, begins with a compound consonant sound, *sk*, moves to the single consonant sound, *s*, and then gives us another and different compound sound, *st*. The sounds accompanying the first and last words starting with *s* are hard consonants— hard like the storm—while at another level, the alliteration of *s*'s lends a malevolent, hissing sound to the line. Again we have Shakespeare's artful use of theme and variation.

Here it's important to emphasize something touched on above. It isn't that the Bard obeys an algorithm, a poetic code, containing the single additional command, "Yoke some of a line's alliterative consonants with various companion consonants." Shakespeare's poetic isn't merely slightly more complex than the call to crudely alliterate. His art involves a matrix of techniques that are more complex by orders of magnitude than that of the poetry-by-numbers school. They are techniques that both guide the poet's hand and work only when the poet takes them in hand, artfully selecting where and when and how to use them. The poet both loves and is beloved by the muse, both gives and receives, both discovers and creates.

The subject of Shakespearean poetics is a book and many books unto itself. Shakespeare's use of alliteration alone could fill one. But one other example will suffice to suggest his exquisite sensitivity to his immediate poetic and dramatic context. Consider the spirit of Ariel's enchanting and justly famous song to Prince Ferdinand, newly washed up on the island:

> Full fathom five thy father lies,
> Of his bones are coral made;
> Those are pearls that were his eyes;
> Nothing of him that doth fade,
> But doth suffer a sea-change
> Into something rich and strange.
> Sea-nymphs hourly ring his knell.
> Hark, now I hear them: ding dong bell.[14]

Unlike the alliteration in Miranda's line noted above, the alliteration in this first line is all theme, without a whiff of variation by way of accompanying consonant sounds. This opening *f* sound occurs four times in the line, each time without an accompanying consonant sound, and thrice in the first three words. And as if this weren't enough, the line's only two nonalliterating words form a slant rhyme. Superficially at least, this has far more in common with the opening line of doggerel about the big, bold, bored dog.

The ear, however, hears the difference. The *b* sound in the doggy poem beats against the ear bombastically, and the effect is out of step with a poem about not a barking, bombastic dog, but rather one suffering from a primitive case of ennui. Ariel's alliteration, in contrast, well suits the situation. The gentler *f* sound imitates the rhythm of the waves unfurling on the island shore. Note also the first line of Ariel's song must quickly establish a poetic form and tone distinct from the conversational verse around it. Ariel, a spirit, is calling us into a world of musical enchantment, of exquisitely wrought artifice. With iambic pentameter both before and after the song, here is no place for a too-subtle poetic architecture.

With this in mind, though, notice that the alliteration of the four *f*'s is not left to sag under its own none-too-subtle and undifferentiated weight. The four alliterating words alternate one syllable and two syllables; and beginning with the fourth word, assonance (the repetition of a vowel sound) emerges to begin tugging against the alliterative pattern. The word *thy* is piv-

---

[14]Shakespeare *The Tempest* 1.2.395-402.

otal: it both disrupts the alliteration and establishes the emerging pattern of repeated long *i* sounds. In the fifth word, the alliterative *f* reasserts itself, coupled to an abortive assonance involving the letter *a*. One might say this is nonexistent assonance, since the *a* sounds in *fathom* and *father* are distinct, but the common letter, even where only seen, together with the kinship of the two vowel sounds, creates a ghostly linkage appropriate to a song confronting the young Ferdinand with the apparent gulf between him and his seemingly deceased father.

As with the doggerel, this line contains an internal slant rhyme, *thy* and *lies;* but here the rhyming word is beautifully apt. Its meaning is precise to the context, playing out the tension between the fading *f* sound and the emerging pattern of long *i*s; and finally, *lies* puns on a second appropriate meaning and, hence, a double-meaning—*lie* as in *to deceive*—since Ariel is, to put it bluntly, lying: Ferdinand's father is alive and well and Ariel knows it.

When college students are confronted with the artifice of blank verse, alliteration, assonance, internal rhyme and other metrical devices, they often balk, thinking, *Please, there's no way the poet was thinking about all that.* This is true—the poet thinks of all this no more than a great basketball player thinks about the precise angle of his body, the height of his jump, his particular backward momentum or the instant of release during a fading jump shot in a game. The player has so developed his craft that he could make the countless little decisions instantaneously by practiced habit and so appears to make them without effort. A given line of Shakespeare may or may not have come to the Bard with such ease: the neoclassical picture of him as an untutored force of nature notwithstanding, Shakespeare undoubtedly thought quite hard and consciously about a great deal in his plays. However, we can understand this without falling into the equally incredible view that every line involved the conscious juggling of a legion of poetic rules or laws. As a student of verse and drama, as a lover of the English word, Shakespeare absorbed the million little intricacies of his mother tongue into his very blood. What we can be certain of is this: great poetry does not happen without effort, without great intellectual work.

As we noted, this insight into the creative process bears on the essentially romantic ideas of many biologists in their effort to explain the emergence of life as a mere unfolding of lawlike processes, a flowering free of the intrusive input of intellect. On this model, the physical constants governing matter and energy work as a kind of endlessly creative algorithm. This is materialism leavened with a dollop of pantheism, where the laws are clocklike gods. But just as the notion of a formula for Shakespeare fails, so too does the idea of an algorithm for life fall apart among the rich and noncompressible details of the living organism.[15]

Return now to Miranda's opening speech, for there is much more to see. As we soon learn, Miranda was taken from Milan when she was three and has lived with her good father on the island for twelve years; she is like clay her father could form without the complexities and contagions of civilization. In her first speech, Miranda's innocence and compassion are revealed in seed form, and as the play unfolds, Shakespeare artfully brings to light, through both her words and actions, the various elements of her character. And each new revelation allows us to better appreciate just how well crafted her opening speech is.

Initially, she appears to be the hoped-for result of all those who believe that human nature is essentially good and who would try to build a heaven on earth by peopling it with such innocence devoid of the ill effects of civilization. She represents what we might call a kind of Pelagian primitivism: the belief (over against Christian orthodoxy) that humans are indeed perfectible by their own efforts, given the right external conditions. But Shakespeare, in the rest of the drama, provides a wider context that calls into question so simplistic a view of human nature. Is it really civilization that corrupts humans, or does the corruption come from within the human heart? Is virtue natural in the sense of being a kind of presocial, original goodness, or is virtue possible only within civilization? If we could create a society from scratch and carefully watch over its every detail, could we eliminate evil and perfect the goodness of its members? If so, under what conditions would it be possible?

Since Shakespeare hadn't the advantage of a Marxist indoctrination at an American or European university in the latter half of the twentieth century, he carried the old-fashioned notion (if *The Tempest* is any indication) that civilization is, necessarily, a hierarchically complex organization of relationships, each position carrying its own set of privileges and responsibilities. The good-hearted but naive counselor Gonzalo seems somehow to have

---

[15]For why the regularities of chemistry and physics cannot account for the aperiodic, nonrandom, noncompressible information found in DNA (apart from the additional problems we discuss in chap. 8), see Stephen Meyer, "DNA and the Origin of Life: Information, Specification and Explanation," in *Darwinism, Design and Public Education*, ed. John Angus Campbell and Stephen C. Meyer (East Lansing: Michigan State University Press, 2004), pp. 248-62; this is available online at <http://www.discovery.org/articleFiles/PDFs/DNAPerspectives.pdf>.

gotten a taste of a contemporary, secular education, for he waxes eloquent about the utopia he would create were he king of the island, a realm without weapon or marriage or ruler. The worldly wise Antonio (Prospero's usurping brother) and Sebastian (who is plotting to murder his own brother, King Alonso, and seize the throne) listen to Gonzalo's utopian fantasy and then effectively cut it to ribbons,[16] noting among other things that Gonzalo's decision to banish all authority from his fantasy kingdom would banish Gonzalo's very authority to banish all authority.[17] Shakespeare understands that there's no question of whether power will be wielded in a society. In the power vacuum of naive communism, the wicked will seize authority and dystopia will supplant utopia. The questions are, who will wield power, how should they wield it, and how and to what degree can freedom and order coexist?

Our progress through the play allows us to see even more in its beginning. By the end of act 2 we can return to Miranda's opening speech and discover there another layer of artfulness. The play, it's becoming apparent, is among other things a meditation on power and its proper use, and the seeds of it were in Miranda's opening lines. We see that she is, simultaneously, repulsed by her father's use of power and eager to wield power to sink "the sea within the earth" so as to order the situation as she would have it.

Now, to say the least, sinking the Mediterranean Sea in a fit of compassion could not remain an isolated act. Obviously it would have consequences far beyond what Miranda had intended, upsetting the balance of nature in a host of unforeseen ways. A generous reading would take her words as hyperbole: she wishes her father had never conjured the horrible tempest. Live and let live. But it soon becomes clear that her words, if not to be taken literally, are by no means empty words. She wants to cancel her father's terrible use of power, but to do so demands power—the magical power she has seen her father exercise many times during their life on the island. But does she yet know enough about human nature to make such a decision wisely? Prospero's power, we soon learn, is the means not only of getting them off the island, but of bringing King Alonso to repentance and Miranda into a happy and prosperous marriage. And this happens *through* the tempest he conjured, although Miranda is not yet wise enough to understand.

---

[16]Shakespeare *The Tempest* 2.1.143-65.

[17]For other utopian imagery of the day see also Francis Bacon's *New Atlantis*, Sir Thomas More's *Utopia* and Michel Montaigne's *Of the Cannibals*.

The play offers several instances of Prospero's wisdom over against the faulty judgment of those who come under his authority. Consider Prospero's treatment of the noble Prince Ferdinand. Upon witnessing his daughter and the prince falling in love at first sight, Prospero decides that "this swift business / I must uneasy make, lest too light winning / make the prize light," so he subjects Ferdinand to a variety of labors.[18] Some critics have seen this as a thinly veiled excuse for sadism. But as the play reveals, these critics lack the wisdom about human nature that guides Prospero, for the labor testifies to the positive effects of hardship on human character. Ferdinand's love for Miranda grows rather than withers under the labors given to him. Obedient to Prospero, he is blessed rather than harmed.

At the same time, other elements of the play make clear that Miranda's criticism of her father's use of power, even if partial and ignorant of certain highly relevant facts, has some merit. Prospero's magic ultimately sets all sorts of things right on the island, and portraying him as a villain rather than as a complex but largely sympathetic hero clearly misses the mark. Nevertheless, Shakespeare studied human nature closely enough to understand the truth in the later aphorism about power's power to corrupt.[19] We see the danger not in Prospero's bare threats to torture Caliban. The creature admits he wishes he could have raped Miranda repeatedly and peopled the island with Calibans. This is a fellow who needs to be scared straight. Rather, we see the corrupting influence of unchecked power in the way Prospero describes the threatened torture. He limns it in loving detail, and not just the better to cajole his recalcitrant underlings. No, at times he seems to wallow in the particulars of it like a pig in the mire. The reason isn't difficult to grasp. Prospero is a man all but mastered by the desire for revenge; and anyone who would stand in his way, even Ariel, a figure for whom he appears to have some genuine affection, becomes a target for his hot rage.

Prospero is no formulaic hero. He is himself is aware that something sick is at work in his spirit; he explains to his daughter that back in Milan he had been too enamored of magic and had turned his administrative duties over to his brother, Antonio, so that he could retire to his ivory tower and pursue a world of magic free from the messy complexities of administration. He wanted to probe the mysteries of power without the accompanying duties

---

[18]Shakespeare *The Tempest* 1.2.448-50.
[19]The aphorism originated with English historian Lord John Acton (1834-1902): "Power tends to corrupt, and absolute power tends to corrupt absolutely."

that come with power; he wanted to live in a private world where he could order things precisely as he wished. This, he tells Miranda, was a key reason he lost his dukedom to his conniving brother.

The play offers other details supporting such a reading. Before entering into this intimate conversation with his daughter about their past life in Milan, Prospero sets aside his magic cloak, as if the robe were somehow unworthy of such an intimate and familial moment (1.2.24). Then at the end of the play he releases Ariel and sets aside his magic robe and staff for good, as if such unchecked power was not fit for the rhythms and institutions of civilization. Thus we see that whereas Miranda lacked the information and shrewdness to understand her father's need to invoke a tempest, she also possessed a deeper wisdom that taught her to fear what unlimited power could do to her father. A being of unlimited virtue could handle unlimited power, but fallen humanity never could.

We are led, then, to deeper considerations not just of humanity, but of divinity as well. "The sky, it seems, would pour down stinking pitch" (1.2.3), Miranda says of the tempest. It's an image we associate with divine judgment—with the destruction of Sodom and Gomorrah—and one in turn reused later in Scripture to describe the fate of the damned. Prospero is playing God, which is always dangerous. The burning pitch is checked only because, as Miranda says, "the sea, mounting to th' welkin's cheek, / Dashes the fire out" (1.2.4-5). This, too, is a description of the storm's fierceness, but it opposes the infernal imagery of the previous line not only literally but also on a symbolic level. In Scripture, water is a source of cleansing and grace, most obviously in the form of baptism but also, as in the case of Jonah, even when it comes as part of a fierce storm. Water and grace deliver Jonah from his foolish retreat from God. And grace, too, delivers Alonso and Prospero. It delivers Alonso to the island where he, in turn, is delivered from his sin. And not only does water deliver Prospero from the island, but by allowing a reunion with his brother, it also delivers Prospero both from the suffocating desire for revenge and from the spiritual danger of virtually unchecked power as absolute ruler of the island. So again, we see how Miranda's language artfully opens onto the work as a whole.

A final example of the art packed into Miranda's opening lines will suffice: the very first word of Miranda's speech. By itself, *if* is nearly but not quite meaningless. But in relationship to the various layers of complexity in *The Tempest*, it becomes more and more meaningful, and we come to understand more fully why that particular *if* exists where it exists. The *if* is a

sign that Miranda does not know whether the tempest she views is the result
of nature, and hence incapable of change, or whether it is one more fantastic
event conjured up by her father. She is the student of her father in many
things, but obviously he has chosen not to reveal his larger plans nor to
teach Miranda his magic; for otherwise she would not be in a state of ignorance about the cause of the storm and would not begin her first speech
with *if.* In short, we first meet the innocent Miranda *wondering* what is going on.

It is no accident, then, that Shakespeare chose for her the name Miranda,
which means a wonderful woman (the feminine of the Latin adjective
mirandus, from the verb *mirari,* "to wonder or to be astonished at"). She is
wonderful, not only in the sense of being a wonder to behold (because of
her beauty and purity), but even more so in being full of wonder, both in
regard to being in a state of ignorance and astonishment brought on by the
tempest and, as we have seen, in regard to the actual complexities of human
nature.

There is more art in Miranda's opening passage and in its relationship to
the larger architecture of the play, but this suffices as a taste of Shakespeare's
genius and as a demonstration of the commonsensical point that her passage
is an integral part of a larger, more complex whole. Even the first word—a
mere *if*—two letters that even a monkey might find on a keyboard, strike us
with a great richness of meaning in the context of the entire drama.

This complex whole, the entire drama itself, is the proper compass both
of our full understanding of the passage's meaning and of our judgment of
Shakespeare's genius. That is, to take up our argument from the previous
chapter, if we were to judge whether this passage could have been generated merely by chance—by hyperassiduous monkeys, by some random
letter-generating machine or by Shakespeare's mind ultimately defined as
the product of natural selection and his thought determined by its material
conditions—we would have to judge it not in isolation, but as an integral
part of the entire drama.

The parts of Miranda's speech—whether it be her entire opening speech,
the first line of that speech or even the first few words—exist in embedded
layers of integration: sentences within speeches, speeches within scenes,
scenes within acts and acts within the play, each serving its role within the
larger whole. An important sign of this relationship is that the more we study
the entire play, the more meaning filled we find any particular passage and
the more appreciative we are of Shakespeare's art.

This depth of meaning, which we find not only in this, but in nearly every passage of *The Tempest,* has led generation after generation to assess it as a work of surpassing genius. We have only touched on its richness. If there were more space, we could enumerate and trace its innumerable puns and motifs and its artful nesting of narratives within narratives alluding to other narratives, all arranged in such a way that, as literary theorist Barbara Mowat argues, lends a surprising plausibility to the cascade of happy events at the end of the tale.[20]

Now is clear the point of studying Shakespeare rather than merely arguing over the abstract probability of randomly generating one of his plays. It is not just that monkeys could not generate Shakespeare's *Tempest* by pecking away at typewriters for millions of billions of years on million of billions of Earths; even more humbling, few if any of us—using our native talents and intelligence, and given the same time as the prodigal monkeys—could produce anything approaching the genius of Shakespeare's *Tempest.*

Also, since we intend to show that nature is not only meaningful but ingenious, we need a good description of what constitutes a work of genius, as well as some grasp of what can and cannot generate such a work. Our exploration of Shakespeare will allow us to consider why his best plays are considered clear works of genius. This in turn will provide working criteria of genius as we consider the artistry of nature. The list that follows is not meant to be exhaustive or definitive or to provide some formulaic and overly tidy "mechanism" of genius detection. And the first criterion we will consider is perhaps the furthest of the criteria from the narrow tidiness of reductionism, a quality many a high school English student knows and fears—*depth.* A person can dig and dig in Shakespeare and keep finding new meanings. It's magnificently dense, rich and subtle. As a friend said about another literary text, "You can just wallow around all day in it and not run out of things to think about."

A library's worth of books have been written about the works of Shakespeare. We see the depth, too, in Shakespeare's vocabulary. He employed one of the largest of any literary author. There apparently weren't enough English words for his needs, so he coined a good many more, perhaps over seventeen hundred. A short sample of a long list: *academe, assassin, be-*

---

[20]Barbara Mowat, "The Tempest: A Modern Perspective," in William Shakespeare. *The Tempest,* ed. Barbara Mowat and Paul Werstine. Folger Shakespeare Library Series (New York: Washington Square Press, 1994), p. 189.

smirch, hobnob, lackluster, madcap, metamorphize, mountaineer, pedant, swagger, zany and *eyeball.*[21]

Of course, a lot of things are deep. The center of the Earth is deep, but we can't get to it. If someone meant for us to get there any time soon, he muffed it. The artistic genius doesn't just bury treasures. She wants her hidden riches discovered, and to that end she lends to her work *clarity,* the second of our criteria for recognizing works of genius. Our need to wade through four hundred years of change in the English language obscures the reality that Shakespeare is rarely ever obscure from mere error or ineptitude. Nor is he obscure for the sake of obscurity. Shakespeare set about to be watched and understood, and succeeded to the point that even uneducated peasants loved him. Shakespeare could have made his plays easier to read, but that's not the same thing. Clarity (from the Latin *clarus,* "bright, shining, illustrious and evident") bespeaks a brilliance that both dazzles and illumines. As such, clarity is the opposite not only of obscurity, but also of drab and easy newspaper prose.

Recall Miranda's speech and our dull paraphrase. The paraphrase is easier to read. But almost nothing is brought before our eyes. The brightness, all of the vividness of scene and emotion, is gone. There are subtleties, too, that have dropped from sight (or rather, no longer exist to be seen). Clarity is not an abstract thing isolated from play and audience. Here the clarity skillfully directs the meanings of the play to English-speaking humans and illumines the depths of the play. It is, moreover, what we might call an anthropic clarity (from the Greek *anthrōpos,* meaning human being), one bent on impressing on the human viewer as much of the play's richness as possible and, indeed, teaching him about the depths and contours of his own human nature.

And here the term *impress* is used very deliberately rather than *communicate. Communication* has been so drained in the modern mind of its rich associations that it says nothing of how deeply a message penetrates.[22] The

---

[21]Some of these, of course, may have been slang that Shakespeare merely picked up and used, but the unusual number of new words introduced into the language through Shakespeare, at a time when London possessed a variety of respected playwrights and poets, suggests that many of these newly penned terms were indeed Shakespeare's own. For an exhaustive list see Stanley Malless and Jeffrey McQuain, *Coined by Shakespeare: Words and Meanings First Penned by the Bard* (Springfield, Mass.: Mirriam-Webster, 1998).

[22]At its root the word *communicate* is full of rich associations (e.g., common, commune, community, Holy Communion). The poverty of the modern word is a mark of a larger cultural shift, brought on by the nihilistic attack on word and Word.

poet doesn't seek merely to pass along data, a drab stream of information. Emily Dickinson said, "If I feel physically as if the top of my head were taken off, I know that is poetry."[23] The easy line is rarely the most brilliant. To borrow his own words, Shakespeare held "the world in awe."[24] He did not always hold its hand.

Augustine discussed this in relationship to Scripture—how the Bible contains a particular teaching often in two forms: one difficult, the other easy. Jesus, for instance, couched some of his teachings in difficult parables and explained to his disciples that the difficulty was intentional. Augustine surmised that the easier passages were to encourage the hungry while the difficult passages were there "because those who do not seek because they have what they wish at once frequently become indolent in disdain" and because "indifference is an evil."[25] The one in love does not pursue indolently, and the lovely woman plays hard to get for a reason.

Miranda was too innocent of human nature to understand this, so her father made Ferdinand's way difficult for a time. Now and again Shakespeare does the same with us: he makes the work of comprehension, as Prospero says of Miranda's courtship, "uneasy . . . lest too light winning / Make the prize light."[26] At other times it is the very strangeness of the thing Shakespeare holds before us—something extraordinary he has discovered that he wants us to discover too—that makes comprehension difficult. Somewhere in his imagination he discovered a complicated prince named Hamlet. We are still getting to know him.

Other times the thing Shakespeare holds before us is too familiar, and Shakespeare would have it grow "rich and strange" for us again, would break down "the automatism of perception" to use Boris Eichenbaum's phrase.[27] To say that the skull was the skull of Yorick does not impress. There was once a man there—body and soul one—a living, breathing human being full, as Hamlet says, of "infinite jest." Before Shakespeare is done with the subject in the midst of the gravedigger scene, Caesar's dust has plugged a hole in a cottage wall and the dust of Alexander the Great has

---

[23]Quoted in Thomas Wentworth Higginson, "Emily Dickinson's Letters," *The Atlantic Monthly* 68, no. 4 (October 1891): 444-56.

[24]Shakespeare *Hamlet* 5.1.199.

[25]Augustine *On Christian Doctrine* 108-13, in *Critical Theory Since Plato*, rev. ed., ed. Hazard Adams (New York: Harcourt Brace Jovanovich, 1992), p. 112.

[26]Shakespeare *Tempest* 1.2.448-50.

[27]Boris Eichenbaum, "The Theory of the 'Formal Method,'" in *Critical Theory Since Plato*, rev. ed., ed. Hazard Adams (New York: Harcourt Brace Jovanovich, 1992), p. 806.

stopped up a beer barrel. Horatio says this is "to consider too curiously," but Shakespeare knows better. He is teaching us, through Hamlet's meditation, to number our days aright.[28]

Shakespeare's genius is both in the threads (the "quotable quotes"), and in the fabric as a whole. A work may be rich but little more than what Henry James described as a "wonderful mass of life," insight without architecture, depth without a governing design.[29] In contrast, Shakespeare brings the most disparate elements into *harmony,* the third of our four criteria for recognizing a work of genius. At his best, nothing is merely tacked on.

Here Coleridge's distinction between fancy and imagination is instructive. "Fancy," he writes, "has no other counters to play with, but fixities." The latter sticks two or more things together to create a monster—say a horse with the head of a man—and the reader will look in vain for a probing intermingling of horse and human personalities in a character so formed. In contrast, imagination "dissolves, diffuses, dissipates, in order to re-create; or where this process is rendered impossible, yet still at all events it struggles to idealize and to unify."[30] Coleridge felt that Shakespeare possessed this imaginative faculty to the highest degree. *Hamlet,* for instance, is a tragedy that is also full of comedy, and not just in the funny scenes. The tragic tone interpenetrates the comical tone and vice versa. And yet the tragedy is not diluted by the comedy but rather deepened and heightened by it. The gravedigger scene is here instructive. The tragic and the comic do not merely tolerate each other but are wedded.

Thus we move from harmony to its cousin, *elegance,* our fourth element of genius. With harmony the emphasis is on the relationship among the parts, a diversity brought into happy association. Elegance emphasizes the unity in diversity. When harmony is discussed, two or more particular elements are often in view, whereas when we speak of elegance, typically we have in view some thing, some intimately ordered whole. Albert Einstein's theory of relativity is elegant because it describes so much so compactly. The elegance of Shakespeare is of a different sort, different from the elegance of the sonnet or even of a Sophoclean tragedy with its severity of line and strict adherence to

---

[28]Shakespeare *Hamlet* 5.1.168-92.

[29]Henry James is here speaking of Leo Tolstoy. Henry James, "Ivan Turgenev (1818-1883)," in *Library of the World's Best Literature,* ed. Charles Dudley Warner (New York: International Society, 1897) <http://www.eldritchpress.org/ist/hjames1.htm>.

[30]Samuel Taylor Coleridge, "Biographia Literaria," in *The Oxford Authors: Samuel Taylor Coleridge* (New York: Oxford University Press, 1985), p. 313.

the unities of time and place. It may be the quality of which *Hamlet* least partakes, but that's a point easily exaggerated. The elegance of great drama, Shakespeare in particular, is not that of the chambered nautilus but of the fawn, not that of the pillars of a Greek temple but of a gothic cathedral. It is the elegance not of the mathematical algorithm but of the living whole.

Now we are aware that twentieth-century philosophy has flung a wealth of ink at the notion of organic wholes. In literary studies, the best deconstructionists expose readings that are guilty of a narrow tidiness. They hold up to the light readings that masquerade as totalizing assessments of works that are really far richer. But the deconstructionist program goes well beyond this. Their common drumbeat is not that artistic works are organic wholes but that instead they contain contradictions and gaps. The man of letters used to call these paradoxes and mysteries, but now we hear *"Aporia, aporia!"* They've found, to quote Umberto Eco, "the little logical termite which nibbles away at it and spoils its perfect self-sufficiency."[31] Their argument stretches well beyond literature. As we will explore in a later chapter, materialism and its heirs ultimately are attacking not merely organic wholes in art but the very notion of an organism—the very notion that sheep and roses and butterflies actually exist first and foremost as sheep and roses and butterflies rather than as a mere conglomerations of subatomic particles. The program is reductionism; the motive, relativism: the insistence that every traditional vision of reality is merely subjective, ready to collapse when pressed at its weakest point.

But the reductionist strategy itself collapses when pressed. Did any of the leading critics in the Western critical tradition ever claim that a literary work formed a totalizing and airtight description of reality? No, that's a straw man, and there's an ironic aptness in its being such. Straw men, scarecrows, do not live in vital relation to their environment. A real man, a real lion, a real work of art, the organism we call the cell—all of these do. We are not claiming for either *Hamlet* or *The Tempest* or any other work of genius the decon-

---

[31] Umberto Eco, *The Aesthetics of Thomas Aquinas*, trans. Hugh Bredin (Cambridge, Mass.: Cambridge University Press, 1988), p. 202. Eco wrote the body of this book before he fell into deconstruction, but he added a preface and an eighth and final chapter decades later. He asserts that Thomas Aquinas's theory of beauty collapses from a logical contradiction, but Eco seems to accomplish the collapse by refusing to recognize the notion of degrees of beauty and degrees of appreciation of beauty. Thus, by Aquinas's logic, the human or artificial work of art is comprehensible but poor in comparison to the living works of art in nature. The artistic creations of God, in contrast, are so rich and complex that we cannot fathom or comprehend them. Voila! Eco declares, contradiction and collapse. But did Aquinas ever claim that the apprehension of beauty had to be a perfect apprehension of perfect beauty? Eco needs to reread Dante.

structionists' straw-man notion of wholeness, nor are we claiming that we wholly understand any particular work of art. Quite the opposite. Together the qualities of genius noted above—*depth, clarity, harmony* and *elegance*—again and again offer to the careful reader the experience of surprise. If one snoops around in Shakespeare long enough, one comes to expect surprises—delightful parallels, double meanings, illuminating motifs and connections of every sort—enough to keep university scholars happily occupied for generation upon generation.

Here an intriguing parallel presents itself between the book of Shakespeare and what Dennis Danielson calls *The Book of the Cosmos*. The readers most likely to ferret out the surprises afforded by Shakespeare are those who most suspect there is some great underlying richness and order that an ingenious author wove into his plays for us to find. In later chapters we shall see that it was just this kind of suspicion that drove the founders of modern science.[32] Christian apologist C. S. Lewis titled one of his books *Surprised by Joy*. It isn't about Shakespeare or science, but it could have been. Each provides, again and again, the joy of surprise.

Consider a final example. The Bard was not, it seems, crippled by false modesty. A careful look at the closing act of *The Tempest* makes this clear enough. The play was apparently Shakespeare's last complete dramatic effort, and the Bard was more than skillful enough to allow Prospero to remain wholly Prospero while using the magician's final speeches to place an exclamation point on his own incomparable career. Consider this soliloquy from the first scene of the final act, long understood to carry autobiographical overtones:

> . . . I have bedimmed
> The noontide sun, called forth the mutinous winds,
> And 'twixt the green sea and the azured vault
> Set roaring war; to the dread rattling thunder
> Have I given fire, and rifted Jove's stout oak
> With his own bolt: the strong-based promontory
> Have I made shake, and by the spurs plucked up
> The pine and cedar; graves at my command
> Have waked their sleepers, oped, and let 'em forth
> By my so potent art. (5.1.41-50)

---

[32]See also Dennis Danielson, ed., *The Books of the Cosmos* (Cambridge: Perseus/Helix Books, 2000).

At first reading, it seems to be about Prospero's use of magic, and of course it is. But it is also Shakespeare speaking about his own dramatic art—not just the technician's art of making a great storm appear on the stage of the Globe Theater, but the dramatist's art of making storms appear in the imagination, storms that set off the audience's reflections about power, sin, virtue, providence and politics.

Shakespeare as Prospero, a benevolent and wise guide on this little island, Earth. Why did he cease using his art, hang up his magic cloak and retire? What connections are there between magic and drama, between the power of the magician and the power of the dramatist? What dangers lurk there? What temptations? The end makes us go back to the beginning of *The Tempest* to plumb its depths again, the better to understand this new level of meaning.

Such is the power of Shakespeare's carefully contrived surprise. We thought, at the end, that we were at the end, that we had understood the "point" of the drama and were now in familiar territory. But familiarity is more likely to breed contempt than humility and wonder, and so the artist must undo it. Coleridge (and later Victor Shklovsky) wrote about the artist's role in defamiliarizing the world around us, in estranging what has become so familiar that we can hardly see it anymore.[33] In sharp contrast, philosophical materialism speeds the work of dulling our vision, rendering the wonder of life as a shuffling of atoms, utterly clocklike, predictable. One thinks of Shakespeare's Macbeth: he killed his king and seized the throne, setting himself against the order of nature and nature's God. In the end he saw all of life as nothing more than "a tale told by an idiot, signifying nothing." Materialism ends in just such madness by strangling life through a logic too-tightly knit.

There was a tearjerker from the early 1980s, *Somewhere in Time,* in which Christopher Reeve plays a man who uses self-hypnosis techniques to travel back to the early twentieth century to pursue an infatuation with a beautiful young actress played by Jane Seymour. At a critical moment late in the film,

---

[33]Victor Shklovsky, "Art as Technique," trans. Lee T. Lemon and Marion J. Reis, *Critical Theory Since Plato,* rev. ed., ed. Hazard Adams (New York: Harcourt Brace Jovanovich, 1992), pp. 750-59. See also "A Report by J. P. Collier of a Lecture Given by Coleridge, 1811-1812," in *Coleridge's Writings on Shakespeare,* ed. Terence Hawkes (New York: G. P. Putnam's Sons, 1959), pp. 43-44. Coleridge writes, "The poet is not only the man made to solve the riddle of the universe, but he is also the man who feels where it is not solved. What is old and worn-out, not in itself, but from the dimness of the intellectual eye, produced by worldly passions and pursuits, he makes new."

Reeve's character inadvertently draws from his pocket a penny from his own time. The penny and the date on it don't belong in the world of 1912 and immediately his rich, meaningful world dissolves and he is swept back to the dreary present. The situation of man under materialism is precisely the reverse. If he can train his eye on a penny that is obviously from another world—can truly see some shining fact that doesn't fit within his reductive, materialist framework—then his seeming world will lose its power to enthrall. In the film, the modern penny expels the protagonist from an older time full of beauty and returns him to a contemporary landscape starved of significance. In our case, the coin of genius rescues us from a landscape drained of meaning and points us toward the real world, toward a place saturated with meaning and worth.

Shakespeare's art is an art of flesh and blood, but it doesn't belong in the world of materialism; it isn't reducible to material causes. And in his art we hear an echo of something materialism would have us forget—the genius of nature, a something more than matter informing matter. To this we now turn.

4

THE GEOMETRY OF GENIUS

The most incomprehensible thing about the

universe is that it is comprehensible.

Albert Einstein

WE NOW TURN FROM THE GLORIES OF WILLIAM SHAKESPEARE to the glories of mathematics. For those of us whose mathematical training happily ended at the battle of long division, it may come as a surprise that the study of mathematics eventually leads beyond the rote learning of mathematical facts to an inner beauty, rigor and harmony that the human intellect finds deeply satisfying. More surprising still—particularly for those trained in the philosophy of materialism—is the revelation that mathematics proves to be such an effective instrument in deciphering the order of nature.

Why should there be any connection at all between the highly abstract world of mathematics and the very concrete world of our everyday experience? It is a curious fact, one that doesn't fit neatly into the worldview of those who believe the universe is a meaningless tug of matter and energy. Recall that physicist Steven Weinberg asserted, first, that the universe "seems pointless" and that we are "a more-or-less farcical outcome of a chain of accidents reaching back to the first three minutes [after the big bang]" and, second, that modern scientists "build telescopes and satellites and accelerators, and sit at their desks for endless hours working out the meaning of the data they gather." This drive, he concludes, "is one of the very few things that lifts human life a little above the level of farce, and gives it some of the grace of tragedy."[1]

As we noted before, these two assertions don't belong together in any

---

[1] Steven Weinberg. *The First Three Minutes: A Modern View of the Origin of the Universe* (New York: Basic, 1977), pp. 154-55.

coherent worldview, much less back-to-back on the concluding page of a physicist's book about the meaning (or nonmeaning) of the cosmos. Since many scientists and philosophers would agree with Weinberg, it is well worth exploring the matter of meaning and meaninglessness in greater depth. How can the universe be both pointless and intelligible? Why are humans—whose intelligence is allegedly the outcome of both natural and sexual selection—not content, like the rest of the animals, to confine their thoughts to ruminations about the daily affairs of animal life? What is the point of studying mathematics if we are only creatures driven by our creaturely drives? Further, what is the point of using mathematics to study a pointless universe? And again, why is mathematics so effective in working out the meaning of the data in our efforts to understand the universe?

Let us begin with the first question: What is the point of studying mathematics? The question was posed long ago to Euclid, an ancient geometer who could properly be called the Shakespeare of mathematics. In one of the few surviving biographical fragments about him, we have an answer to this question. After having learned the first geometrical theorem, a pupil inquired of Euclid, "But what shall I *get* by learning these things?" Euclid called one of his slaves. "Give him a coin," Euclid ordered, "since he must make a gain out of what he learns."[2] Unfortunately, we do not have recorded what effect Euclid's stinging words had upon the student, so we do not know whether the student blushed from embarrassment or was simply stunned by incomprehension. Either way, the point of Euclid's remark is that the study of geometry is intrinsically good and needs no further justification. While it may have practical uses, these are accidental to its true merit, the peculiarly human joy of gaining knowledge about mathematical things. Flipping the student a coin was a way to chastise him, marking him as one with an attitude unworthy of the study of mathematics for its own sake.

Of course, the utilitarian approach to learning rebuked by Euclid is not confined to students of long ago nor to the study of mathematics. Many humanities professors, no doubt, wish they had the candor of a Euclid when confronted by budding utilitarians impatient to know, "What is this good for?" They mean, of course, how is the study of Mozart or art history or Shakespeare going to get them higher-paying jobs? At this point, we professors usually cave and begin babbling about how corporations want more

---

[2]A short biography, including this story, is found in the introduction to *The Thirteen Books of Euclid's Elements,* 2nd ed., trans. Sir Thomas Heath (New York: Dover, 1956), 1:3.

than just number crunchers. "They want people who can think," we reply, "and Shakespeare will help you to think. Ergo, Shakespeare → $hake$peare." To speak thus, we realize, is to sell the Bard for small change. Better to reach into a bag of coins and toss one to the student: "Here, young man, take this if you must make a gain from reading *Hamlet*."

Why go on about this? Because the attempt to justify the teaching of Euclid or Shakespeare in terms of material gain is not far removed from the attempt to reduce the works of Shakespeare or the works of Euclid to some material cause. A Marxist treatment of Shakespeare reduces his genius to the modes of economic production regnant in late-sixteenth century England; a utilitarian justification for reading Shakespeare tallies his worth in terms of the enhancement of economic production today. In both cases, the assumption is that the achievements of our intellect are to be understood only in terms of providing for our bodily self-preservation, comfort and gratification.

The same thing occurs where Darwinism attempts to explain human achievement in mathematics. Why is the human mind, alone among the animals, capable of geometrical understanding? A Darwinist must reply that the successful study of geometry is ultimately the result of selection pressures; that is, it comes from the direct advantage such enhanced intellectual capacity has for the struggle to survive and mate. To be fair, there is some justification for this utilitarian view. One might argue, for example, that geometry came into being first of all as a practical science (in Egypt); that is, it was originally directed to measuring and building and only later became primarily theoretical in Greece. As Cicero famously remarked, the ever-practical Romans would have no part in the merely theoretical study of geometry but confined their interest and labors to its purely practical applications.[3]

Cicero was not praising his fellow Romans for being practical but lamenting their intractable utilitarianism and wishing they would act more like the earlier civilization of the Greeks. He himself had received an excellent Greek education, and therefore was immersed in geometry as taught by those who believed (as Euclid did) that the study of the truths of geometry needed no other justification but was a worthwhile endeavor in itself. If he were transported to a modern-day geometry class, his lament would surely not cease. Any math teacher will assure you that today there are thirty Romans to every Greek in introductory mathematics.

Despite this Roman or practical side of geometry, the Darwinian account

---

[3]Cicero *Tusculan Disputations* 1.5.

of the development of human intelligence does not explain the extraordinary intellectual gap between the capacity to reason geometrically in regard to mere survival, the far more extraordinary capacity entailed in purely theoretical geometry such as that taught in Euclid's *Elements,* or even more, in the kind of mathematics used in contemporary physics. To imagine that Darwinian selection mechanisms could have seized upon a series of small but immediately beneficial genetic variations to produce a species capable of producing a Newton or an Einstein works no better than explaining Shakespeare by such means. "We have certain skills—for example, we can jump streams and catch falling apples—which are necessary for getting by in the world," notes physicist Paul Davies, "but, why is it that we also have the ability to discern, for example, what's going on inside atoms or inside black holes? These are completely outside the domain of everyday experience . . . not at all necessary for good Darwinian survival."[4]

We can understand how slightly faster cheetahs would have a selective advantage in chasing down gazelles; but if we discovered a distinct species of cheetahs that could run 6,000 mph rather than 60 mph, far faster than any gazelle, we would have to look elsewhere for an explanation. This is the situation where we encounter mathematical genius. Charles Darwin's way around such excessive capacities was to put them down to indirect natural selection, that is, sexual selection. Here, of course, we aren't far removed from Freudian theory, in that the highest achievements of the intellect are ultimately reducible, as effects, to sexual desire.[5]

What is interesting about all this, however, is the incongruity that nearly always exists in those who persist in trying to explain away the amazing capacities of the human intellect by such reductionism. Surely, nearly all teaching biologists, even and especially advocates of Darwinism, have felt the twinge of revulsion when asked by students to justify the study of living things in terms of some material gain. Insofar as they react this way it is because the teachers regard the study of biology as needing no other justification than the pure intellectual delight of knowing the truth about biological

---

[4]From the science film *The Privileged Planet: The Search for Purpose in the Universe* (La Habra, Calif.: Illustra Media, 2004). Paul Davies develops this and other points related to the surprising discoverability of the laws and constants of physics in *The Mind of God: The Scientific Basis for a Rational World* (New York: Simon and Schuster, 1992). The book by Guillermo Gonzalez and Jay W. Richards, *The Privileged Planet* (Washington, D.C.: Regnery, 2004), develops the argument even further, particularly in chaps. 6 and 10.

[5]Although in contrast to Freudian theory, Darwinism links the presence of sexual desire to the capacity for procreation, rather than to the mere desire for pleasure itself.

things. At the same time, Darwinism as a theory seeks to reduce the intellectual capacity for such study to just the kind of "gain" that the utilitarian student desires—some advantage for self-preservation, either direct or indirect.

How does this all relate to Weinberg? His account of the universe is, of course, merely Darwinism writ large; that is, the cause of the order we experience—even the laws of nature—is ultimately the production not of intelligence, but of chance. The chain of accidents stretches from the first three minutes, all the way through biological evolution on our obscure little planet, and right up to the capacities of the scientist patiently trying to ferret out the meaning of the data in an effort to achieve a deeper understanding for its own sake. Mathematics is essential for science. The "telescopes and satellites and accelerators" couldn't even be built without mathematics. But to the question, "Why should we work out the meaning of the data, especially if it takes endless hours sitting at desks?" Weinberg gives an eminently nonutilitarian answer: "The effort to understand the universe is one of the very few things that lifts human life a little above the level of farce, and gives it some of the grace of tragedy." And what specifically is the grace bestowed? "The more the universe seems comprehensible," Weinberg asserts, "the more it also seems pointless."[6]

Contra Weinberg, that isn't tragedy. It's farce, a blind lark in the theater of the absurd. The philosophers tell us life is meaningless. The scientists invest meaninglessness with scraps of meaning through mathematics; they slave away, studying formulas and peering to the far reaches of the universe trying to uncover the hidden order of things, only to discover a sign at the edge of the universe announcing, "Hail nada full of nada."[7]

But perhaps if we look more closely, we will see a way out of the absurdity. Hidden in Weinberg's words are two very important, connected truths, one about human nature and one about the universe. The truth about human nature is that humans take immense joy in knowing for its own sake. Indeed, the scholar's intellectual exhilaration often increases the further the object is removed from considerations of our animal existence. The exhilaration the young physics lover feels when she finally gets it, when she understands how energy equals mass times the velocity of light squared, doesn't include concern for her next meal. She may even forget to eat. One

---

[6]Weinberg, *First Three Minutes*, p. 154.

[7]The phrase is taken from Ernest Hemingway's nihilistic café worker in "A Clean, Well-Lighted Place," *The Short Stories of Ernest Hemingway* (New York: Collier, 1986), pp. 379-83.

of a scientist's greatest intellectual joys occurs when she finds that a formula or figure quite effectively illuminates reality, particularly when, like an unearned gift, it sheds light in an unanticipated direction. This peculiar joy is entirely unconnected with our animal needs; it wells up from somewhere deeper and stretches far above our earthly existence.

Such an experience may seem unachievable, given that few of us are likely to grasp anything from higher physics or advanced mathematics, but the experience can also be had at an elementary level with Euclid's *Elements*. This is particularly so thanks to the fact that Euclid combines, in one person, the genius of the greatest geometer and the pedagogical genius of the greatest teachers. And so just as we asked readers to experience Shakespeare, we now ask them to leap into Euclid. Those of us who are not mathematically inclined need not fear. Euclid is a most gentle teacher, and he rarely fails to awaken a deep love of mathematics even in the most mathematically phobic. Euclid did his work about three centuries before the birth of Christ. Historically, he was indebted to a line of mathematical geniuses reaching back to the sixth-century mathematician and mystic Pythagoras. In Euclid's justly famous *Elements,* he was able to bring together a long line of geometric achievements, improve and clarify many of the existing proofs, provide new and superior proofs, and organize the various parts of geometry and mathematics into a most beautifully well-ordered and extraordinarily accessible whole. Not only has the *Elements* been studied for over two millennia, but if we trace our own achievements in science, we find that modern science, in large part, stands on the shoulders of Euclid, for many of the watershed discoveries between the seventeenth and nineteenth centuries were made possible by the application of Euclidean geometry to nature.

There are so many things in the *Elements* worthy of our attention, but here we will focus on just one of them, a most ingenious proof of the most famous geometrical truth, the Pythagorean theorem: On any right triangle, the square on the hypotenuse equals the sum of the squares on the other two sides of the triangle. This is represented as $a^2 + b^2 = c^2$. Unfortunately, most children today are merely taught the formula rather than the demonstration. But it is only in learning the demonstration that we can truly be said to know the truth of it, for it is only then that we know why it is necessarily so—that is, why it is true in all possible cases. We teach our students to parrot math formulas and then wonder why more of them don't fall in love with mathematics. Euclid had the solution more than 2,000 years ago: show, don't tell. Mold a novice into a knower. Even more important for our argument, it is precisely in knowing that we experience the peculiarly human intellectual joy and know it to be fundamentally disconnected from any relationship to our self-preservation or sexuality.

As we might have expected, it was not Pythagoras who gave us the most ingenious proof of the Pythagorean theorem, but Euclid.[8] Pythagoras's proof was based on a theory of proportions, which was not only more advanced than would be appropriate for a beginner, but also only as good as the theory of proportions itself (which was, apparently, somewhat defective at the time). Euclid provided a much simpler and far more powerful proof. As such, one doesn't need to know much to follow his explanation, even though it contains within it innumerable hidden connections to later propositions and geometrical mysteries. Given a definition and a couple of propositions (illustrated below), the learner is off and running.

To build up to the proof in regard to right triangles, we need first to study the parallelogram. A parallelogram isn't one-dimensional, like a line, or three-dimensional, like a box. It's a type of two-dimensional figure. To put the matter in colloquial terms, it's flat, as flat as can be. But there are lots of two-dimensional shapes—circles, triangles and on and on. As seen in the illustration below, parallelograms are four-sided figures where both pairs of opposite sides are parallel and of equal length. A rectangle is one type of parallelogram; a square is another.

From consideration of the parallelogram, we arrive at two important demonstrations illustrated by the figures below: first, "parallelograms which are on the same base and in the same parallels are equal to one another" (prop. 1.35); and second, if a parallelogram and a triangle share the same base, the parallelogram's area is double the triangle's (prop. 1.41).[9]

---

[8] As Proclus said, "If we listen to those who like to record antiquities, we shall find them attributing this theorem to Pythagoras and saying that he sacrificed an ox on its discovery. For my part, though I marvel at those who first noted the truth of this theorem, I admire more the author of the *Elements,* not only for the very lucid proof by which he made it fast, but also because in the sixth book [prop. 31], he laid hold of a theorem even more general than this and secured it by irrefutable scientific arguments. For in that book he proves generally that in right-angled triangles the figure on the side that subtends the right angle is equal to the similar and similarly drawn figures on the sides that contain the right angle." Proclus *A Commentary on the First Book of Euclid's Elements,* trans. Glenn Morrow (Princeton, N.J.: Princeton University Press, 1970), 3.426.47, pp. 337-38. In this analysis of Euclid we will be using the three-volume Heath translation cited above.

[9] The abbreviations herein for propositions in Euclid's *Elements* give the book and proposition numbers (e.g., proposition 1.41 is proposition 41 in book 1).

Proposition 1, 35 and 41

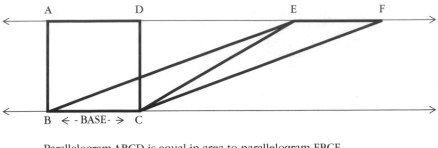

Parallelogram ABCD is equal in area to parallelogram EBCF.
Triangle BCE is 1/2 the area of parallelgram ABCD.

Figure 4.1.

From these truths about parallelograms, we move to the famous proof of
the right triangle. Euclid's proof is so beautiful that we must savor it, step by
step. And again, we ask readers to adopt the view of Euclid, and not that of
his utilitarian student. Don't worry about what the proof is good for (not
even what it is "good for" in relationship to the argument of this book).
Throw yourselves into the proof, and experience his genius just as we tried
to experience the genius of Shakespeare. We begin simply by drawing a
right triangle (i.e., a triangle in which one of the angles measures ninety degrees) with the hypotenuse down.

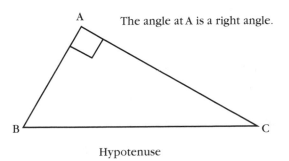

Figure 4.2.

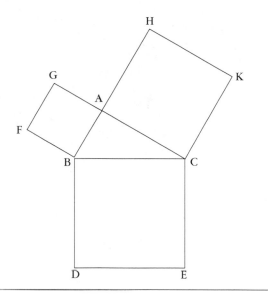

Figure 4.3.

Then we draw squares on each of the three sides.

Next we draw a line from point *A* to point *L*, parallel to line *BD*, running from the vertex of the triangle to the very bottom of the square on the hypotenuse. This parallel line, as we shall see, is one of the most amazing strokes of genius in all geometry.

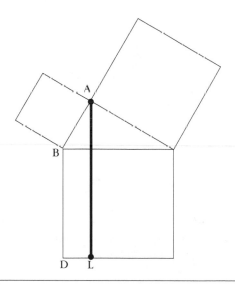

Figure 4.4.

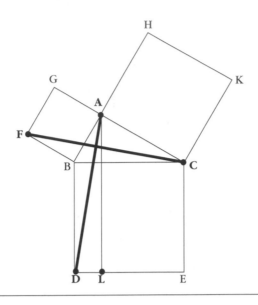

Figure 4.5.

We now connect point *F* with point *C,* and point *A* with point *D*. From this figure, the entire proof is made.

How? As we walk through it, keep referring back to the illustrations. Because we constructed squares on the sides of the right triangle, we know that angles *ABF* and *CBD* are right angles; and obviously, since they are both right angles, they are equal. (See fig. 4.6 below.) Add the angle *ABC* to each. Since the same angle is being added to two equal angles, the results are equal; that is, angle *FBC* is equal to angle *ABD*. But each of the triangles, *FBC* and *ABD,* is using a side from each square. Therefore, the two triangles have an equal angle in between two equal sides.

If we then lay one of these triangles down upon the other—the equal side *FB* on top of *AB,* the equal angle *FBC* on top of the equal angle *ABD,* and the equal side *BC* on top of *BD*—we find out that they perfectly coincide. (Imagine placing a nail at point *B* and then rotating triangle *FBC* clockwise until *FB* rests right on top of *AB*.) Thus, the remaining side of each is equal and the two remaining angles are also equal. Therefore, the triangles *FBC* and *ABD* are equal.

Now comes the geometrical marvel. Because *FGAB* is a square, lines *GA* and *FB* are parallel. Since line *AC* is a continuation of *GA,* the entire line *GC* is parallel to *FB*.[10]

---

[10]We're being a bit quick here. Euclid proves this through proposition 1.14.

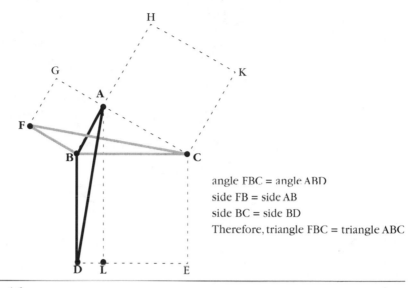

angle FBC = angle ABD
side FB = side AB
side BC = side BD
Therefore, triangle FBC = triangle ABC

Figure 4.6.

The interesting thing is that the square *FGAB* and the triangle *FBC* share the same base (line *FB*), and are under the same parallel (line *GC*). But as we've already seen, "if a parallelogram have the same base with a triangle and be in the same parallels, the parallelogram is double of the triangle" (prop. 1.41). Thus, the square *FGAB* is double (in area) the triangle *FBC*.

Of course, *AL* was drawn parallel to *BD*, and triangle *ABD* is on the same base

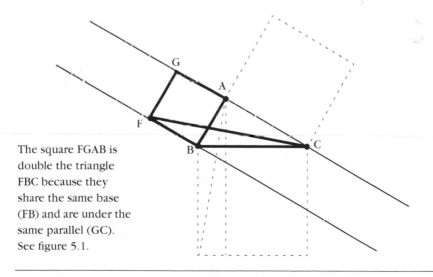

The square FGAB is double the triangle FBC because they share the same base (FB) and are under the same parallel (GC). See figure 5.1.

Figure 4.7.

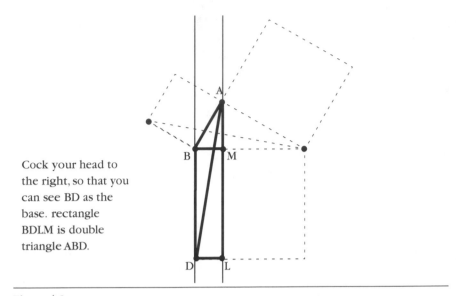

Cock your head to
the right, so that you
can see BD as the
base. rectangle
BDLM is double
triangle ABD.

Figure 4.8.

(line *BD*) and under the same parallel (line *AL*) as rectangle *BDLM*. Therefore, we find the same relationship, where rectangle *BDLM* is double triangle *ABD*.

But since we have already shown the two triangles *FBC* and *ABD* to be equal, then certainly the doubles of equals will be equal. Thus the square *FGAB* is equal in area to the rectangle *BDLM*. If we repeat the entire procedure for the square on the other side, we find that the square *AHKC* is equal to the rectangle *LECM*.

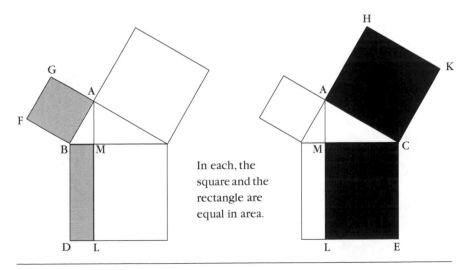

In each, the
square and the
rectangle are
equal in area.

Figure 4.9.

Three Right Triangles

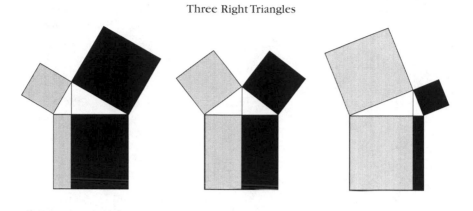

Figure 4.10.

And so we have the proof. The two rectangles (*BDLM* and *LECM*) added together obviously make up the square on the hypotenuse *BC* (i.e., the square *BCED*), and they do so for every possible right triangle. Thus, we can set *any* right triangle on its hypotenuse, draw the squares on each side of the triangle, drop a vertical line from the vertex of the triangle straight down, perpendicular to the hypotenuse, and that line will divide the square on the hypotenuse into two rectangles equal to the squares on their respective sides.

Since these are always formed by cutting the square drawn on the hypotenuse into two rectangles, and the rectangles are always equal to the two squares on the two non-hypotenuse sides of the right triangle, we have our universal proof: For *any* right triangle, the square on the hypotenuse is equal to the sum of the squares on the other two sides.

If the elegance of the proof hasn't yet seized you, go take a look at the older proof Pythagoras developed available in any history of mathematics. Unlike Pythagoras's proof, Euclid's is simple enough to be grasped by beginning geometry students who have not yet worked up to an understanding of proportions. Euclid places this proof in book 1 of the *Elements,* but he doesn't take up the more difficult notion of proportionality until book 5. When he finally does apply proportionality to the right triangle in book 6, proposition 31, Euclid proves that any similarly described figures—not just squares, but *any* similar figures, no matter how wild[11]—will have the same

---

[11] This includes circles as well, since he proves in prop. 12.2 that "circles are to one another as the squares on the diameters."

"Pythagorean" relationship when they are placed on the sides of a right triangle: that is, the figure on the hypotenuse will equal the sum of the similar figures on the other two sides of the right triangle. Thus Euclid surpasses all previous proofs by virtue of the universality of his demonstration. Such is the genius of Euclid.

Let us reflect on this geometric proposition and, more importantly, on our experience of knowing. First of all, in regard to attempts at reductionism, we might well ask why Euclid should bother trying to surpass Pythagoras. We suspect that even Marx himself would be hard pressed to explain the intellectual leap in terms of some fundamental change in the modes of production governing Pythagoras and Euclid respectively. Might it make sense in light of Freud's Oedipus complex? Was Euclid suffering from the "anxiety of influence"? Was he an oedipal geometer seeking to overthrow his intellectual father, Pythagoras? Or could we somehow, in the lines of the diagram of book 1, proposition 47, discover the lines of anxiety caused by Euclid's fear of death? Or perhaps the proof is simply evidence of some trait elaboration connected to sexual selection?

If we are honest and unhindered by such reductionist presuppositions, we will reject these attempts to reduce the glory of Euclid to such ignoble and wildly unlikely causes. Instead, we would simply and sanely say that there seem to be two causes of the proof. On the one hand, there is a power inherent in human reason, the drive to understand, not for any ulterior motive, but purely for the sake of knowing; and this drive takes us far beyond the practical realm as we search for knowledge that is universal, true for all cases and true for all times. Surely, for all merely practical purposes, neither Pythagoras's nor Euclid's proof was necessary; as carpenters knew long before Pythagoras, a knock-together wooden triangle with side lengths of 3, 4 and 5 gives you a perfect right angle every time.

The other cause of such a proof is outside of us, we might say. It is the inherent, necessary and universal order of geometrical things themselves, an order that preexists our attempts to uncover it and that inflames the geometer's desire to know it. The ultimate reason that we cannot reduce either Euclid's genius or Euclid's proof to some materialist, reductionist cause is that the proof concerns truths independent of any such cause. The ultimate cause of the truth lies in the nature of right triangles. This state of affairs didn't evolve. Such mathematical truth has an inner necessity and order that exist outside the realm of chance and independently of human intelligence. The human intellect, whatever its cause, is strangely fit to grasp it, and that fitness cannot be reduced to the survival of the fittest.

There is something else weird about a right triangle. Like the circle, it's one of the sturdiest things in the cosmos and, at the same time, utterly immaterial. Physical circles are never really perfectly circular; physical right triangles always fall short of perfection. We grasp such geometrical figures only with our minds, never with our hands. When we do grasp the demonstration, even though we may do it through particular drawings, we lay hold of why it must be so in any particular case from the nature of the right triangle as such—but there is no "right triangle as such" out there floating in the cosmos that we can either see with our eyes or grasp with our hands. The struggle to understand the demonstration about right triangles is the struggle to grasp something that is immaterial.

But if we are successful in this struggle, and we glimpse the necessity of the geometric relationships as such, then and only then can we see why it must be so in every possible case and not just in this or that particular drawn triangle. Even more profound, if we reflect on our own reflection, we receive a more beautiful proof, a demonstration that we have, in our reason, a power to grasp immaterial truths—a power that somehow exceeds the particular, physically defined powers of our senses and imagination and is capable of grasping universal truth. Could this be a proof of the immateriality of the soul?

Perhaps it is. True observation in science does not twist or distort what it sees and understands to make the evidence fit into a preconceived intellectual framework. True observation tries to see a thing as it exists and acts, even and especially when we are that "thing." In the act of knowing we observe in ourselves what a human is. In knowing, we know ourselves to be rational animals—not disembodied intellects or intellects trapped in bodies, and not mere animals driven by hunger, thirst, comfort and sex.

But let us reflect even more deeply on our experience of working through Euclid's proof. What is it in humans that takes such delight in geometry? Human beings are pattern-seeking, beauty-loving creatures and, consequently, they find delight in combining and repeating geometrical figures (as the continual use of such figures throughout the ages in art and architecture attests). This love of beauty shows itself even more fully in the very human desire to understand the underlying order of geometrical figures. And it accounts for the steady stream of Euclid enthusiasts for more than two thousand years. Edna St. Vincent Millay (1892-1950) embodied her ardor in a poem:

Euclid alone has looked on Beauty bare.
Let all who prate of Beauty hold their peace,
And lay them prone upon the earth and cease
To ponder on themselves, the while they stare
At nothing, intricately drawn nowhere
In shapes of shifting lineage; let geese
Gabble and hiss, but heroes seek release
From dusty bondage into luminous air.
O blinding hour, O holy, terrible day,
When first the shaft into his vision shone
Of light anatomized! Euclid alone
Has looked on Beauty bare. Fortunate they
Who, though once only and then but far away,
Have heard her massive sandal set on stone.

This near-rapturous paean to Euclid is not just the result of an aesthetic appreciation of geometrical figures—something that occurs on the surface level; even more, it is praise for the union of intellectual truth and elegance in Euclid's presentation. The beauty in Euclid goes all the way down. In the *Elements,* as in all works of genius, it seems as if two things, truth and beauty, somehow unnaturally separated, have been reforged into a fresh unity by the fire of Euclid's extraordinary intellect.

Why do we delight in the beauty of geometrical figures and proofs, an appreciation obviously not related to material need? What did Darwin say? First, he argued that "the sense of beauty obviously depends on the nature of the mind, irrespective of any real quality in the admired object" so that "the idea of what is beautiful, is not innate or unalterable." Second, he argued, "If beautiful objects had been created solely for man's gratification," then we should find "less beauty on the face of the earth" before humans appeared. But since beautiful objects in nature, such as the "beautiful volute and cone shells of the Eocene epoch [55 to 38 million years ago]" existed long before human beings, then it is absurd to believe that they were "created [so] that man might ages afterward admire them in his cabinet." Third, although the ultimate origin of the appreciation of beauty—"the reception of a peculiar kind of pleasure from certain colors, forms, and sounds"—is "a very obscure subject," the sense of beauty is not particular to human beings but occurs both "in the mind of man and of the lower animals."[12]

---

[12]Charles Darwin, "Utilitarian Doctrine, How Far True: Beauty, How Acquired," chap. 6 of *The Origin of Species* (New York: Mentor, 1958), pp. 188-90.

But on all three points Darwin's theory fails to account for the kind of beauty that human beings experience in mathematics. To respond in order, in regard to Euclid's propositions, the sense of beauty obviously does depend, not only on the nature of the mind, but also and primarily on real qualities of the admired object, the geometrical figure. That is, the "Beauty bare" to which Millay referred is, again, the union of truth and beauty in the admired geometrical object, and it is this union that Euclid's demonstration so elegantly makes manifest.

Second, certainly the internal order of geometrical figures existed long before the Eocene period, even though no human being either existed or had discovered them. It's tendentious at best to argue that the elegance of book 1, proposition 47 wasn't real until it was discovered three centuries before the birth of Christ. That would be akin to arguing that Albert Einstein's equations—universally regarded by physicists to be of the utmost elegance—were not really beautiful until produced in the early twentieth century. What makes Einstein's equations beautiful is not only that the equations are elegant, but that they express certain aspects of the order of nature mathematically—an order that existed long before and independently of any human attempt to grasp it. It's perhaps true that if there were no one to regard or know about these mathematical truths, they would not be known as beautiful. But against even this a wisdom older and deeper whispers to us through the fog of materialism: those mathematical truths were regarded as beautiful for as long as they have been, longer than *we* have been.

Third, while some animals do apparently appreciate beauty, only humans can grasp the elegance of a mathematical demonstration and (even more amazingly) discover geometrical or mathematical relationships that exist in the order of nature. Since the intrinsic elegance of the order of nature refers not only to the beauty of butterflies and plants (which conceivably could be related to natural and sexual selection), but also to the larger order of nature encompassing nonliving things, there clearly exists intrinsic elegance that cannot be reduced to natural and sexual selection.

We have spent some time focusing on beauty because our appreciation of mathematical beauty extends to the most abstract intellectual realms, including those inhabited by theoretical physicists. We may now ask a crucial but frequently overlooked question: What *right* have we to expect that our human capacity for mathematical abstraction and our human appreciation of elegance would yield any knowledge of nature? If, after all, the universe itself were randomly produced and did not have us in mind, and if our own reasoning capacities and love of beauty were likewise randomly produced, could we reasonably expect mathematics to be an effective tool for us in "working out the meaning of the data"?

This is the very question asked by the famous physicist Eugene Wigner in his equally famous rumination, "The Unreasonable Effectiveness of Mathematics in the Natural Sciences."[13] In his oft-quoted words,

> The enormous usefulness of mathematics in the natural sciences is something bordering on the mysterious. . . . There is no rational explanation for it. . . . The miracle of the appropriateness of the language of mathematics for the formulation of the laws of physics is a wonderful gift which we neither understand nor deserve.[14]

We are so used to the association of mathematics with science that the strangeness of the effectiveness of mathematics doesn't properly awaken our wonder. We must rouse from our wonder-less slumber to grasp how marvelous it is to find meaning in the data through mathematics.

To do so, we must grow more childlike in our capacity for wonder. It is rare that either mathematicians or physicists have a sufficient philosophical bent to marvel at the origin of numbers and the rather strange ability to count things. One such mathematician is R. W. Hamming:

> I have tried, with little success, to get some of my friends to understand my amazement that the abstraction of integers for counting is both possible and useful. Is it not remarkable that 6 sheep plus 7 sheep make 13 sheep; that 6 stones plus 7 stones make 13 stones? Is it not a miracle that the universe is so constructed that such a simple abstraction as a number is possible? To me this is one of the strongest examples of the unreasonable effectiveness of mathematics. Indeed, I find it both strange and unexplainable.[15]

Of course, Hamming is consciously playing off Wigner's essay; but more than this, he is pushing the analysis back to the beginning, back to the original abstraction. Mathematics has a history; that is, humans are not born knowing arithmetic, algebra, calculus or geometry, let alone the extraordinary develop-

---

[13]"The Unreasonable Effectiveness of Mathematics in the Natural Sciences" was originally published in *Communications in Pure and Applied Mathematics*, 13, no. 1 (February 1960): 1-14. We are citing the article as reprinted in *Symmetries and Reflections: Scientific Essays of Eugene P. Wigner* (Woodbridge, Conn.: Ox Bow Press, 1979), pp. 222-37.

[14]Wigner, *Symmetries and Reflections,* pp. 222, 237.

[15]R. W. Hamming, "The Unreasonable Effectiveness of Mathematics," *American Mathematics Monthly* 87 (February 1980): pp. 81-90.

ments of mathematical systems in the last two centuries. Numbering, as Hamming notes, begins on a very rudimentary level by counting very ordinary and obvious things. By what right would we expect that the numbers so derived would be applicable on any other level of reality? Why should the universe be such that whole numbers—originally derived from counting quite ordinary, visible objects—are so effective in helping us discern the hidden orders of nature? What relationship is there among the 2 in two sheep, the 2 in two stones, and the 2 in the equation $v = 2\pi r/T$ or $e = mc^2$ or $a^2 + b^2 = c^2$? With Hamming, we should find it "both strange and unexplainable" that this number, any number, should apply so effectively at so many diverse levels of reality, both above and below our scale of vision. How odd that the human discipline of mathematics is so meaningful, given its humble beginnings.

The same could be said of geometry. Certainly the notions of a circle or a sphere were originally abstracted from round objects, just as the notions of a point from a dot and a line from something straight and slender. By imitating these objects in art, adornment and architecture, humans made them available for more abstract analysis such as occurs among the Pythagoreans or in Euclid. Yet from Ptolemy to Copernicus, Galileo, Johannes Kepler and Newton, Euclid's almost self-contained reflection on the beauty and truth of geometry—self-contained in that the advance from one demonstration to the next remains in the abstract realm of geometry—became the textbook for scientific discovery, especially in regard to astronomy, though in other areas as well.[16] How strange that geometry should prove so effective in illuminating nature, given that it originally arose from a quite simple abstraction from ordinary objects, and further that it was developed theoretically, almost entirely independently of any worry about its being effective in illuminating nature.

That "strangeness" is the heart of Wigner's wonder at the "unreasonable effectiveness of mathematics in the natural sciences." As Wigner points out, the strangeness lies in the mysterious and surprising connection between the nonutilitarian human desire for formal beauty that drives mathematical development and the order of nature existing independently of our intellect. While it is "unquestionably true that the concepts of elementary mathematics and particularly elementary geometry were formulated to describe entities directly suggested by the actual world" (e.g., two sheep suggesting the number 2, the moon suggesting a circle), "the same does not seem to be true of

---

[16]Along with Euclid's *Elements*, the work of Apollonius of Perga on conic sections also proved extraordinarily fruitful.

the more advanced concepts, in particular the concepts which play such an important role in physics." As he explains, "Most more advanced mathematical concepts, such as complex numbers, algebras, linear operators, Borel sets—and the list could be continued almost indefinitely—were so devised" not in reaction to some obvious feature of nature but as "apt subjects on which the mathematician can demonstrate his ingenuity and sense of formal beauty."[17] Of this ingenuity, Wigner remarks, "certainly it is hard to believe that our reasoning power was brought, by Darwin's process of natural selection, to the perfection which it seems to possess."[18]

The very development of mathematics, Wigner asserts, whether in ancient Greece or in the last two centuries, depends on this peculiarly human desire for order and beauty that permit "ingenious logical operations" that "appeal to our aesthetic sense both as operations and also in their results of great generality and simplicity."[19] What is so amazing that it appears "unreasonable" is that such purely abstract systems should not only be roughly applicable to aspects of the natural order, but upon closer inspection and elaboration, yield undreamed of precision and unexpected discovery beyond the original application.

Wigner notes several examples. The geometry of the parabola was developed two centuries before Christ by the Greek Apollonius; it was then used by Newton in the seventeenth century to decipher parabolic motion in his formulation of the law of gravitation. Matrices were developed in the nineteenth century, and then in the twentieth they found a use in elementary quantum mechanics. And the Lamb shift developed by Hans Bethe and Julian Schwinger as a purely mathematical theory in the twentieth century was later used in quantum electrodynamics.[20] In each case, the applicability of some purely mathematical system, developed for its own sake according to the mathematician's desire for formal beauty, stretched far beyond any reasonable explanation of merely accidental resemblances to nature, so that mathematical beauty led to truths of discovery. The mathematical formulas or figures turned out to be more meaningful than their founders ever suspected; or in Wigner's words, "we 'got something

---

[17] Wigner, *Symmetries and Reflections,* pp. 224-25. As Wigner notes (p. 225), he is indebted to philosopher-chemist Michael Polanyi's work on the importance of our aesthetic desire or, as Polanyi calls it, "a passion for intellectual beauty." See Michael Polanyi, *Personal Knowledge* (Chicago: University of Chicago Press, 1962), pp. 187-190.
[18] Wigner, *Symmetries and Reflections,* p. 224.
[19] Ibid., p. 225.
[20] Ibid., pp. 231-33.

out' of the equations that we did not put in."[21]

That does not mean—we hasten to add—that every aesthetically pleasing formal mathematical system is useful for physicists or chemists, or that every seemingly successful empirical application of mathematics stands the test of time. As Wigner points out, "only a fraction of all mathematical concepts is used in physics,"[22] and as should be obvious from even a cursory reading of the history of chemistry, physics and astronomy, successful applications are often overturned by empirically intransigent data. In fact, it is often when the data do not fit the formerly illuminating mathematical formulas that physicists go searching for more powerful mathematical systems. And so, while reality stands in a close relationship to the discipline of mathematics, it is reality that defines the mysterious effectiveness of mathematics and not vice versa.

We may now draw several important conclusions from our analysis of mathematics that will aid us in our exploration of the universe as meaningful. First, no one can reasonably argue that mathematics has not been an effective tool in helping science find the "meaning of the data." For scientists, the greatest and most peculiar intellectual exhilaration occurs when they find that the order of mathematics illuminates the order of reality. This is not a passionless, accountantlike correspondence of lines and ledgers, but a participation in an ethereal union of beauty and truth—the beauty and truth of the mathematical order matching some aspect of the natural order. It is interesting that, as biologists (following Darwin) have become more reductionist in regard to beauty, physicists have come to a new appreciation of the centrality of beauty in regard to the relationship of mathematical equations to reality.[23]

Further, if we look back over the history of science, we find that the effectiveness is layered in a kind of tutorial fashion; that is, in regard to earlier science, relatively simple mathematics was very effective in illuminating significant aspects of nature. Euclidean geometry worked quite well up until nearly the end of the nineteenth century, and it still works quite well for ordinary purposes. Further penetration into the natural order has required more and more complex mathematics, and yet (as Wigner attests) the strange effectiveness still holds. That such layered, tutorial illumination occurs is truly unreasonable if the universe were ordered randomly. We could

---

[21]Ibid., p. 232.

[22]Ibid., p. 229.

[23]See, e.g., Graham Farmelo, *It Must Be Beautiful: Great Equations of Modern Science* (London: Granta, 2002); and Robert Augros and George Stanciu, *The New Story of Science* (Lake Bluff, Ill.: Regnery, 1984).

imagine, with random ordering, that by some mercy of fickle chance, a purely accidental relationship of some mathematical system would "map onto" a particular aspect of nature, but we would never expect it to effectively illuminate the natural order beyond that merely accidental relationship. Yet if we keep finding that multiple mathematical systems "map onto" nature—calling us from one steppingstone of discovery to the next—then it is certainly reasonable to suspect a conspiracy of reasoned order.

Second, we must be careful what is meant by saying that mathematics is effective, for herein lurks an error that has caused much confusion. As far back as the Pythagoreans in sixth century B.C. Greece, the strange effectiveness of mathematics has been an occasion for awe—so much so that the Pythagoreans (in Aristotle's words) thought that "the principles *[archas]* of mathematics were the principles of all being."[24] The Greek word *archē* has a double meaning that the English *principle* doesn't catch. It means literally "the beginning," but a beginning in the sense of the first cause of what follows. Thus, the Pythagoreans were so enamored by numbers and geometrical figures that they took mathematical things to be causal, rather than as abstractions that happened, for some reason, to be effective in illuminating certain regularities in nature. On this view (if we might simplify it to the point of mild parody), the number 1 isn't something we abstract from counting sheep; one, the principle of unity, is the cause of each sheep being one sheep rather than a morass of disconnected parts.

In the Renaissance, the Pythagorean view (via the rise of Neo-Platonism) was tempered by the Christian view of a divine maker. Galileo's statement about the relationship of mathematics to the universe is illustrative:

> Philosophy is written in the great book which is ever before our eyes—I mean the universe—but we cannot understand it if we do not first learn the language and grasp the symbols in which it is written. This book is written in the mathematical language, and the symbols are triangles, circles, and other geometrical figures, without whose help it is impossible to comprehend a single word of it; without which one wanders in vain through a dark labyrinth.[25]

Galileo's view is an advance over that of Pythagoras, for he no longer has mathematicals causing things like the oneness of sheep. On his view, a cosmic Orderer grounds the orderliness Galileo has discovered. The mathemat-

---

[24]Aristotle *Metaphysics* 985b25-26.
[25]Galileo Galilei *Opere Complete di Galileo Galilei* (Firenze, 1842) 4:171. Quoted in E. A. Burtt, *The Metaphysical Foundations of Modern Physical Science* (New York: Doubleday, 1932), p. 75.

icals aren't themselves primary and causal; rather, God has written them into nature as the symbol system that unlocks its mysteries. This modification, of course, came about through Christianity, which asserts that God, not mathematicals, is the cause of nature's order. For Galileo, then, the effectiveness of mathematics is not unreasonable. On the contrary, the effectiveness of mathematics is the result of God's wisdom ordaining an intelligible order and condescending, we might say, to form it according to the "language" of a mathematics that the human intellect is capable of formulating.

Unfortunately, whatever Galileo's intentions, some his followers would seize on the Pythagorean element in his writing and reify it, giving rise to one of the great confusions of modern thought, one that so imbues our understanding today that we will only grasp its significance with great difficulty. It is all too easy, given the successful application of, say, Euclidean geometry, to again believe that numbers and figures are truly causal, that the geometrically defined laws of nature actually cause objects to move or act in conformity with the purely formal mathematical relationships used to enunciate the laws.

It's important, then, to clarify in what sense mathematics is the language of nature. There are wonderful mathematical regularities, but what about when we arrive at the a-periodic and disorderly order of genetic information? This cannot be mapped, much less generated, by a mathematical formula any more than a mathematical formula could be discovered to generate *Hamlet*. The effectiveness of mathematics has its limits. As philosopher Alfred North Whitehead reportedly quipped, "If you think one and one is always two, try adding a match and a stick of dynamite." On a less glib note, Whitehead rightly warned against the "fallacy of misplaced concreteness," the "error of mistaking the abstract for the concrete."[26] The error is proportionately as tempting as one's abstractions are illuminating.

The danger in discovering the great effectiveness in mathematics comes in the urges to cover everything by mathematics and to smother anything that resists—to see mathematical abstractions as more real than what we are using them to illuminate, what is right before our eyes. This is the very plague that afflicts the scholars on the floating islands in Jonathan Swift's *Gulliver's Travels*. They have one eye pointed inward and one upward—leaving no vision for what lies before them:

> Their houses are very ill built, the walls bevel without one right angle in any
> apartment, and this defect arises from the contempt they bear to practical ge-

---

[26] Alfred North Whitehead, *Science and the Modern World* (New York: Free Press, 1953), p. 51.

ometry, which they despise as vulgar and mechanic, those instructions they give being too refined for the intellects of their workmen, which occasions perpetual mistakes. And although they are dexterous enough upon a piece of paper in the management of the rule, the pencil, and the divider, yet in the common actions and behavior of life, I have not seen a more clumsy, awkward, and unhandy people.[27]

Indeed, they are so self-absorbed that anyone of them wishing to be heard brings with him a servant wielding an inflated bladder filled with dried peas. With this the servant, or "flapper," must occasionally give his master and his master's companion a good whack to shake them from their reveries and put the conversation back on track. Swift's parody of people living in the world of abstraction even while stumbling and bumbling through the real world before their very eyes captures an important truth. The world of mathematics is a world of abstraction, a step away from reality, not reality itself. It is *through* mathematics, not *in* mathematics, that scientists find meaning in the data. The data is about reality, about the order of beings in nature. That is why reality always determines whether any particular mathematical formulation is applicable and effective.

Why go on about this point? If we reify mathematics, we soon fall into the error of treating the mathematical relationships expressed in laws as if they had divine causative powers—an absurd if subtle kind of idolatry. Unfortunately, the modern mind has all too happily submitted to such idolatry, practicing a strange kind of numerical self-mystification.

With the enormous effectiveness of Newton's mathematical analysis of nature, for example, the followers of Newton began to believe that only the mathematical was real. But the constriction of reality to the canons of a reified mathematics brought rebellion with its success. The tidy, heroic couplets of the eighteenth century clicked by, one after another. People grew restless. And the Romantics, reacting against the sterile watchmaker view of the cosmos, mounted a rebellion, though it was one without a firm foundation, the ground soggy from what T. E. Hulme would later call "spilt religion."[28] They embraced another idol. Having suffered the wildness taken out of nature by scientists bent on pressing all of nature into a clocklike, rational mathematical order, they began to worship irrational wildness itself and to cast an increasingly suspi-

---

[27]Jonathan Swift, *Gulliver's Travels* (1726), pt. 3, chap. 2. Excerpted in *The Norton Anthology of English Literature*, 4th ed., ed. M. H. Abrams (New York: Norton, 1979), 1:2080-81.

[28]T. E. Hulme, "Romanticism and Classicism," in *Critical Theory Since Plato*, rev. ed., ed. Hazard Adams (1924; reprint, New York: Harcourt Brace Jovanovich, 1992), p. 729.

cious eye upon order. The best Romantic poets managed to balance freedom and order, but the seeds that would lead beyond Walt Whitman to much of the undisciplined free verse of the twentieth century had been planted.

Despite the Romantic protest and its long-term consequences, the attempt to constrict nature unnaturally persisted and took an even stranger form. In the next generation, Darwin saw something else in nature than William Wordsworth's field of lovely daffodils—a daughter dead and nature "red in tooth and claw." There was no God present, he surmised, other than the god of necessity, some fixed and elegant law. His later followers would go further still, forcing the wild beauties of the living world through a mathematical rubric: making nature fit at all costs, reducing it all to numbers, probabilities, mutation rates; and adorning it with vague references to chaos theory and complexity—endless talk of anything but the living breathing organism.

To understand what went wrong we must return again to Galileo's words—the assumption that since mathematics (for him, primarily Euclidean geometry) is so wonderfully effective, that without this "language . . . it is impossible to comprehend a single word of [nature]; without which one wanders in vain through a dark labyrinth." Here the astronomer and physicist—in awe of the mathematical regularities he has uncovered—lets his rhetoric run away from him. He has expressed the marvelous insight that mathematics is a surprisingly effective tool in discerning the order of nature, but it's simply not the case that nature is simply incomprehensible without mathematics—a view among later followers which led to the belief that what cannot be subsumed under mathematical expressions either does not exist or is a mere "subjective" projection on essentially mathematical nature. On this view, the *only* meaningful language is mathematics; and since our everyday language and experience are not governed by mathematics, then our everyday language and experience are not meaningfully related to reality. As a consequence, deep reflections based on our everyday language and experience are taken to be groundless. Out with Shakespeare, Aristotle, St. Thomas Aquinas and any and every dramatist, philosopher or theologian who spoke not in the language of mathematics. In with the mathematicians, the chemists and physicists and, trailing behind them, their various hangers-on, the social scientists, the psychiatrists.

As we've seen, it doesn't end there. The final error is this, one made famous by the German nihilist philosopher Friedrich Nietzsche: having slain the Architect, and with him his designing and governing intelligence, we determine that *we* are really the architects. After all, mathematical systems are purely formal constructions of *our* intellect and imagination, rather than things out there.[29] It is we who impose these artificial creations on the "data" to bring order out of chaos. There really is no meaning in the data and hence no intrinsic meaning in the universe. So the argument goes. This view has grown common enough to invite parody, specifically a bogus physics paper submitted to the journal *Social Text,* from which we now quote:

> The discourse of the scientific community, for all its undeniable value, cannot assert a privileged epistemological status with respect to counter-hegemonic narratives emanating from dissident or marginalized communities. These themes can be traced, despite some differences of emphasis, in Aronowitz's analysis of the cultural fabric that produced quantum mechanics; in Ross' discussion of oppositional discourses in post-quantum science; in Irigaray's and Hayles' exegeses of gender encoding in fluid mechanics; and in Harding's comprehensive critique of the gender ideology underlying the natural sciences in general and physics in particular.[30]

The paper was accepted, and later the author, Alan D. Sokal, announced that the paper was a hoax, including its many spurious textual references. Christopher Pearson and Benjamin Wallace-Wells later explained,

> Sokal believed that deconstructionism was willfully opaque because its ideas, such as hopeless relativism, are impossible to defend in sustained argument. Deconstructionism could only compensate for this deficiency by cloaking its claims with pretentious words for basic concepts.[31]

---

[29]For an extraordinary analysis of the constructive nature of the modern project in regard to mathematics, see David Lachterman, *The Ethics of Geometry: A Genealogy of Modernity* (New York: Routledge, 1989).

[30]Alan D. Sokal, "Transgressing the Boundaries: Towards a Transformative Hermeneutics of Quantum Gravity," *Social Text* 46/47 (spring/summer 1996): 217-52. This is available online at <http://www.physics.nyu.edu/faculty/sokal/transgress_v2/transgress_v2_singlefile.html>. Sokal is or at least was a political leftist who believed that *Social Text*'s brand of reader response-deconstructionist relativism undermined the ability to protest and revolutionize. He's a materialist of the Steven Weinberg variety discussed in chapter one, a person who believes in retail purpose and meaning even as his worldview rests on a materialist foundation of sand in that it creates intractable problems for the possibility of ontological categories like rationality, purpose, and meaning. Sokal's gesture, though potent, was a rear guard action doomed to failure, for he has eschewed the very being who can ground both morality and meaning.

[31]Christopher Pearson and Benjamin Wallace-Wells, "Social Text: Fish's Other Flop," *Dartmouth Review,* February 3, 1999. This is available online at <http://www.dartreview.com/issues/2.3.99/stflop.html>.

Suffocating in the materialist darkness, where all meanings are mere human fabrications, the mind at some point may feel itself thrusting off all of this gloominess, this "hopeless relativism," shocked awake by the *reductio ad absurdum* of mathematical and material reductionism. "But we don't live in a chaos, now do we?" one might summon the courage to say. "We live in a cosmos—messy at times, but Euclid was . . . well, he was onto some *thing*, something more than matter, more than mere imagination."

As we have argued, if the order of nature preexists our attempts to grasp it and, consequently, if the strange effectiveness of mathematics depends on the preexistent order of nature to be effective, then nature is intelligibly and ingeniously ordered. Exemplifying both surprising depth and a stunning harmony and elegance, such ingenious design necessarily implies a designing genius.

As we saw previously, a number of materialists in the academy have sought to dismiss the category of genius, presenting it as merely the pointless outworking of blind forces like natural and sexual selection. But again, this just-so story has its limits. Does anyone really wish to argue that the truth that "the three interior angles of a triangle are equal to two right angles" (prop. 1.32) derives from some sort of preorganic natural selection? What was the probability that it would be the case that in right triangles the squares on the hypotenuses would equal the sum of the squares on the other two sides (prop. 1.47)? The question is absurd. Probability, chance, had nothing to do with it. Rather, a kind of inner necessity and order exist in geometrical figures independently of human intelligence; and human intelligence, whatever its causes, is strangely fit to grasp it. One may retreat to the metaphysics of multiple universes, imagining some other cosmos with a different set of dimensions and geometrical principles, but this explanation, while touching in its faith in the unseen, fails to explain why the geometry of our particular universe is so discoverable and why that geometry initiated a series of accessible steps to Kepler, to Newton, to Einstein and beyond, as if the universe were set up to draw us patiently along as students. Following philosopher Robin Collins's argument, Guillermo Gonzalez and Jay Richards write,

> For some reason the ultimate laws of physics give rise to mathematically simple theoretical laws at each conceptual level, even for those later judged inadequate, such as Newtonian space-time. This odd truth allows each conceptual level to serve as a ladder to the next level. If the theoretical laws could not be simple and yet relatively precise at each conceptual level, we could probably

not discover them for that level, and hence could not progress from level to level toward the fundamental laws of physics.[32]

The more we uncover, the more it looks like there is a conspiracy of order, an idea that allows us to see the genius of Euclid in a different light. He becomes not the founder of geometry, the great master, but an apprentice hurrying after the true Master, the first Geometer, whose work is bright with clarity—rich and strange and elegant, surprising and delighting us with its unexpected harmonies—and all of it oddly fitted to the human mind and imagination. And as we will see in the chapters that follow, that's only the half of it. The first Geometer was something else besides, and the order of nature, while amenable to mathematical analysis, is far richer, having about it the smell of sweat and soil, flesh and blood and fire.

---

[32]Guillermo Gonzalez and Jay Richards, *The Privileged Planet: How Our Place in the Cosmos Is Designed for Discovery* (Washington, D.C.: Regnery, 2004), p. 215.

5

THE PERIODIC TABLE

A Masterpiece of Many Authors

Randomness alone can never produce a significant pattern,
for it consists in the absence of any such pattern.

Michael Polanyi, *Personal Knowledge*

The argument now turns from the mathematical world of abstraction to the world of chemistry and to human beings as knowers of the chemical order, an order illumined by the great literary classic of chemistry: the periodic table of elements. By attending to the history of our discovery of the elements and their order, it will grow increasingly clear how many ways human beings *as knowers* contradict the canons of materialist reductionism. In the next chapter, we will show how the chemical order itself contradicts materialism, so that materialism cannot account for either the knower or the known. Even though we are revealing the ultimate purpose of both chapters on chemistry, we invite readers to assume the same attitude toward the material as they had toward William Shakespeare and Euclid and, first of all, enjoy the knowledge revealed for its own sake.

Like Euclid's *Elements* and Shakespeare's *Hamlet*, the periodic table of elements that hangs on the wall of every chemistry class is a timeless masterpiece. This table, however, isn't the work of one man but of a line of people reaching back to the first philosophers of ancient Greece, and further still to the first artisans of ancient civilization. The table is laid out so neatly on the wall of our high school chemistry classes that it gives the impression that the elements are self-evident letters, an obvious chemical alphabet that was easily decipherable. But that is the result of our seeing the elements primly laid out on a wall chart and without our understanding the long, circuitous road

1 H																	2 He
3 Li	4 Be											5 B	6 C	7 N	8 O	9 F	10 Ne
11 Na	12 Mg											13 Al	14 Si	15 P	16 S	17 Cl	18 Ar
19 K	20 Ca	21 Sc	22 Ti	23 V	24 Cr	25 Mn	26 Fe	27 Co	28 Ni	29 Cu	30 Zn	31 Ga	32 Ge	33 As	34 Se	35 Br	36 Kr
37 Rb	38 Sr	39 Y	40 Zr	41 Nb	42 Mo	43 Tc	44 Ru	45 Rh	46 Pd	47 Ag	48 Cd	49 In	50 Sn	51 Sb	52 Te	53 I	54 Xe
55 Cs	56 Ba	57* La	72 Hf	73 Ta	74 W	75 Re	76 Os	77 Ir	78 Pt	79 Au	80 Hg	81 Tl	82 Pb	83 Bi	84 Po	85 At	86 Rn
87 Fr	88 Ra	89** Ac	104 Rf	105 Db	106 Sg	107 Bh	108 Hs	109 Mt	110 Ds	111 Uuu	112 Uub						

*	58 Ce	59 Pr	60 Nd	61 Pm	62 Sm	63 Eu	64 Gd	65 Tb	66 Dy	67 Ho	68 Er	69 Tm	70 Yb	71 Lu
**	90 Th	91 Pa	92 U	93 Np	94 Pu	95 Am	96 Cm	97 Bk	98 Cf	99 Es	100 Fm	101 Mb	102 No	103 Lr

Figure 5.1.

to the discovery of their nature and ingenious order. While it was a long and winding road, the journey was anything but random, for it was defined by (or better, constantly corrected by) nature itself. If the order were not in nature—and in nature in a way that we could discover it—then the road would have led anywhere and nowhere, and the journey would have been, as the author of Ecclesiastes put it, meaningless, a chasing after the wind. Happily, such was not the case. The periodic table of elements is, in fact, deeply meaningful, illuminating layers of order in the material world.

As wonderfully intelligible as the periodic table of elements is to us now, it would appear to ages past as mere gibberish. Indeed it would appear that way to many of the scientists who contributed to it. The drama of the discovery of the periodic table is much like the drama of a mystery story, wherein the characters are surrounded by clues. They discover those clues bit by bit, but for a long time nobody can see how they all fit together to solve the mystery. For the most part, elements are "hidden" in complex compounds much as the clues of a murder mystery are jumbled together at a crime scene. The elements are not nicely isolated and labeled Fe, Na, Al, P, H, C and so on.

If by analogy the elements were to be considered letters of the alphabet, then it would be accurate to say that in nature we encounter not letters, but almost exclusively spoken words, sentences and, most often of all, elaborately staged theatrical productions. Moreover, even on the levels closest to the chemical elements, the "words," the simplest chemical compounds, form unified substances with their own distinct qualities. For this reason, it was extraordinarily difficult to guess what is or is not elemental. On the level of everyday experience, water does not appear as H_2O, our exhaled breath does not appear as CO_2, and table salt does not appear as $NaCl$.

The analogy to human letters is illustrative, though it should be handled with caution. We have seen it used for the mathematical orderliness of the universe, genetic information and now here with the periodic table. Consider the latter two together; that is, consider compounds, protein strands, cells and complex organisms as, roughly speaking, words, sentences, paragraphs and books, respectively. On this analogy the elements of the periodic table are letters. (They are themselves each much more complicated than letters, but we will bracket that for now.) The chemistry of life is like an unknown alphabet and language rapidly spoken to us. We would have a horrendous time distinguishing its constituent parts, particularly where certain letter combinations create a fundamentally new sound, like *s* and *h* to create *sh*. But beyond distinguishing the letters, we would need to decipher the language's grammar. On this analogy, the periodic table not only lays out the letters but, in its very ordering and in the revealing content of its symbols, suggests the grammar of chemistry.

For those of us who struggled in fear under the stern gaze of our grammar teachers, this analogy might appear daunting, but that's all to the purpose: although it's relatively easy to understand the main lines of the periodic table of elements, it was extraordinarily difficult to put together in the first place—far harder than learning English grammar. For most of human history, both the elements and their order were almost completely unknown, even though the chemical drama was continually played out all around us. To discover the elements we had to penetrate numerous layers of complexity (even to find out that some of the elements were often right under our noses). We humans live on the level of maximum chemical integration and complexity—the level of complex chemical sentences, paragraphs and, most often, elaborate dramas—and there is no way simply to leap over the many layers of integration to the elements to get to the "bottom" of things. Trying to do so almost invariably led to errors and distortions.

So how, then, did we decipher the great drama of nature in regard to the chemical elements? We might say that we were both pulled and pushed into discovery. We human beings cannot remain at a distance from the complexities of nature: if we are not lured in by our desire to understand them and our appreciation of their beauty, then we are pushed in by our need to provide food, clothing, shelter and defense. To begin with the push, humans are born (to use an apt expression from ancient Greek philosophy) "naked, with reason and hands," and we must deal directly with the complexities of nature just to survive. In this sense, necessity is the mother of discovery, for even on this level, we are immersed in the complex drama of chemical existence as we learn the distinct properties of things like fire, wood, stone, earth, water, plants, animals and air.

But even before the necessities are fully met, the lure of beauty blossoms and bears it own fruit. According to renowned historian of metallurgy Cyril Stanley Smith, "aesthetic curiosity" is the real mother of invention, for "most of man's inventions have first appeared in decorative rather than practical applications." He continues,

> Metallurgy began with the making of necklace beads and ornaments in hammered naturally occurring copper long before "useful" knives and weapons

were made. The improvement of metals by alloying and heat treatment as well as most methods of shaping them started in jewelry and sculpture. Casting in complicated moulds began in the manufacture of statuettes. Welding was first used to join parts of bronze sculpture together. . . . Ceramics began with fire-hardening of fertility figurines molded of clay; glass came through attempts to glaze quartz and steatite beads. Most minerals and many organic and inorganic compounds were discovered for use as pigments. Indeed, the first record that man knew of iron and manganese ores is found in the prehistoric cave paintings where these ores provided the glorious reds, browns and blacks.[1]

As Smith makes clear, these earliest of artisans were animated by their fascination for beauty, not utility. Historian of metallurgy Robert Raymond concurs: "All the evidence we have of early metallurgy supports this [Cyril Smith's] opinion. Because of the unusual character and initial rarity of metals, they were first used for decoration rather than utility, for ornaments rather than knives."[2] Thus, as important as our desire for self-preservation is, there would be no periodic table without our very human love of beauty. Elaborating on this point, the great mathematician Henri Poincaré said, "The scientist does not study nature because it is useful to do so. He studies it because he takes pleasure in it; and he takes pleasure in it because it is beautiful. If nature were not beautiful, it would not be worth knowing and life would not be worth living."[3] Beauty draws us beyond the merely practical, acting as a kind of preparation for the love of the elegance of the order of nature itself. As it was with geometry, beauty was, for the founders of modern chemistry, the way to elemental truth.

It is interesting, then, that the element most often exposed in nature in its pure form, in sufficient quantities to be visible, is gold. For this reason, it was probably the first element discovered, long before the category of "element" had itself been discovered. Archaeologists have discovered gold ornaments dating all the way back to the late Neolithic Period, near the end of the fifth millennium B.C. Copper ornaments have been found dating even

---

[1]Cyril Stanley Smith, "Aesthetic Curiosity, the Root of Invention," *Anvil's Ring* (spring 1996), accessed online at <www.pennabilli.org/testi/Smith_EN.htm>. On the importance of beauty see also his excellent essays "Matter versus Materials: A Historical View" and "Art, Technology and Science: Notes on Their Historical Interaction," in *A Search for Structure: Selected Essays on Science, Art and History* (Cambridge, Mass.: MIT Press, 1981), pp. 112-26, 191-241, as well as his amply illustrated *From Art to Science: Seventy-Two Objects Illustrating the Nature of Discovery* (Cambridge, Mass.: MIT Press, 1980).
[2]Robert Raymond, *Out of the Fiery Furnace: The Impact of Metals on the History of Mankind* (University Park: Pennsylvania State University Press, 1986), p. 8.
[3]Henri Poincaré, *The Value of Science* (New York: Dover, 1958), p. 8.

further back, to about 9000 B.C., but it is most likely that the use of gold as a worked metal predated copper because gold occurs in greater concentrations of purer, more visible form. For this reason, historians conjecture that the earliest gold ornaments, because of their obvious beauty and worth, were simply taken up again by later peoples and reworked.[4] Gold made a perfect metal for the earliest artisans because they could pound the malleable stuff into attractive shapes without the aid of fire. The irony, of course, is that its very softness, not to mention its unwieldy density, makes it impractical for use as a weapon or a tool.

To offer a stunning example, the exquisite helmet of Meskalamdug (from Ur, Mesopotamia, c. 2700 B.C.) is entirely decorative, made in the form of a wig with every hair on the prince's head reproduced in flowing gold. The artistry gives evidence of the most advanced metallurgical techniques of beating and annealing, far beyond the rest of the Mesopotamian culture.[5] Yet all of this would not protect the prince against a blow of a sword. The least practical of metals is the most valued, prized because humans are more fundamentally artists than warriors.

This love of beauty that draws artisans ever more fully into the intricacies of nature's chemical order is also clearly evident in other great "practical chemists" of the ancient world: the makers of perfume, pottery and dye. In each case, if artisans had stuck to mere practicality, little or no headway into the penetration of nature's order would have occurred. After all, dyed clothes do not repel wind, rain, snow or a boar's tusk any better than those that remain the natural color of linen, wool or cotton. The addition of perfume serves to enhance physical beauty, not ward off dangers. Plain pots hold water as well as those covered with elaborate artwork.

Now, of course, a Darwinian would likely interject, "Perfume, elaborately colored clothes, glittering jewelry—yes, we are born naked with hands and the bare ability to reason, but that is why we imitate the animals, covering our bodies with the equivalent of beautiful mating plumage and dousing ourselves with artificial pheromones. So, you've plainly lost the argument, for it seems the real and ultimate origin of the science of chemistry is none other than sexual selection."

---

[4]See R. F. Tylecote, *A History of Metallurgy*, 2nd ed. (Brookfield, Vt.: The Institute of Materials, 1992), p. 5; and Robert Raymond, *Out of the Fiery Furnace*, p. 9.

[5]For these examples and an interesting account of the place of gold in civilization, see Carol H. V. Sutherland, *Gold: Its Beauty, Power and Allure*, 3rd ed. (London: Thames and Hudson, 1969), esp. chaps. 1-3.

On the contrary. Our present argument doesn't deny any of the natural reality of human beings; indeed, we are trying to restore it. Humans, as rational animals, do have a desire for self-preservation. We desire food; we desire protection from the elements; and we desire protection from danger. Further, since we are animals that generate by sexual intercourse, male and female humans desire each other sexually for the sake of procreation. To deny or degrade these obvious and wonderful aspects of our being would be both foolish (and, for a Christian, heretical, the heresy being Gnosticism or Manichaeism or, in later forms, Catharism or Albigensianism, all of which denied the goodness of the body). We affirm, then, that beauty is, on some levels, certainly related to sexual attraction. Any fool—and every poet—knows that. What we deny is the crudely dogmatic reduction of the desire for beauty to these levels alone.

And so, we are not trying to ignore the body, as if humans were all head; rather, we object to those who wish, for the sake of their argument, to cut off the head and present a human being as a creature from the neck down (or even from the waist down). Thus, ours is the more inclusive argument, the one that truly describes our entire human appreciation of beauty; it doesn't dogmatically exclude the higher or reduce what is higher to the lower aspects of our nature. Darwin, in contrast, felt compelled to do just that, noting that the existence of beauty for its own sake, or more properly, for the sake of human beings, "would be absolutely fatal to my theory."[6] As it turns out, the development of modern chemistry—which has served (incorrectly) as a foundation for modern materialist reductionism in general and for Darwinism in particular—would not have occurred if not for the very human desire of beauty for its own sake, of beauty sought and shaped simply because it is beautiful.

Thus it was that beauty led to truth. We find the situation, then, growing more and more curious: beauty in nature is both a lure and a guide to truth. Ancient people, too, were enamored of the beauty of truth itself. It is not only the love of beauty for its own sake that separates us from the animals. There is also among us the impractical activity of philosophy, for philosophy primarily seeks the truth of things for the love of truth, just as Euclid believed that geometric truth was desirable for its own sake and not for some extraneous utility. This aspect of the human character was crucial for chem-

---

[6]From "Utilitarian Doctrine, How Far True: Beauty, How Acquired," chap. 6 of Charles Darwin, *The Origin of Species*, 6th ed. (New York: Mentor, 1958), p. 188.

ical discovery. To make significant headway into the secrets of the elements, we had to join the love of gold and things golden to the love of nature's order itself. And indeed, it was this intellectual turn that initiated our search for elements as elements.

In the West, philosophy began with Thales of Miletus (fl. early sixth century B.C.), and since Thales initiated the first profound speculation about the ultimate order of nature, historians of chemistry always begin with Thales's argument that the ultimate element, or principle (in Greek, *archē*), was water. According to Thales, water was the primal element or substance—that "out of which comes every being, and from which all things first are generated and into which they are finally destroyed." Water was the substance, the permanent thing, "that underlies all changes."[7] This assertion is not as strange as it may sound. There can be little doubt that water is absolutely essential to our existence and somehow gives life to all that lives. Even more profound—although Thales could not have known it—is that water is one of the simplest of compounds, consisting of one oxygen atom and two hydrogen atoms, the latter being the very simplest of atoms. The analysis of water into hydrogen and oxygen in the eighteenth century A.D., almost two and a half millennia after Thales, was (as we shall see) the great watershed for modern chemistry. Thus, Thales's conjecture that everything was ultimately made from water was a magnificent, educated first guess—incorrect but wiser than he could have imagined.

Other philosophers soon entered the fray, putting forward different contenders for the title of ultimate element. Thales's contemporary, Anaximander (d. 547 B.C.), also of Miletus, thought that the source of all things was the infinite (or the unlimited or unbounded, in Greek, the *apeiron*). From the infinite come the great "opposites," hot, cold, dry and moist or, more tangibly, fire, earth, air and water. These opposites, when mixed in varying proportions, create all that we experience in the physical world. Sounds a bit strange? Again, if we look around, without any sophisticated modern chemical knowledge, Anaximander's conjecture makes good sense. What seem to be the most fundamental things we actually experience? All living things come from the earth and need water to live, as well as the fire of the sun; and we living things are surrounded by air and filled with breath. Yet, Anaximander was searching for a deeper, truly elemental and perma-

---

[7]This is Aristotle's statement from his *Metaphysics* 983b (translated by Benjamin Wiker). We have nothing directly from Thales himself.

nent source underlying even these obvious "elements" of earth, air, fire and water; and he chose the highly abstract and quasimathematical, quasitheological concept of the infinite. Regardless of the inadequacies of his conjecture, this search for a deeper, permanent source of what appears as changing to our senses is the very essence of science, and the attempt to capture that order in mathematical terms has been essential to scientific progress.

Other ancient philosophers offered rival theories, with different concepts and entities acting as elemental. Anaximanes, a fellow citizen of Miletus and follower of Anaximander, argued that the infinite itself was actually air, for air (as we see in breath) is the cause of life; and for this reason, he considered air to be a life-giving god, with the other three elements coming from air. Air by being rarified became fire; by condensation became water; and by further condensation, earth. Following somewhat the same pattern, Heraclitus of Ephesus (fl. c. 500 B.C.) chose fire as fundamental, the divine cause of order and, hence, the cause of the other elements. When fire is contracted, it becomes moist air; when contracted further, water; and further still, earth. The earth, however, returns to water by liquefaction, and the water to air, and the air to fire, and so the cycle runs eternally.

Others looked to mathematics as somehow elemental. Remember Pythagoras from our earlier discussion of mathematics. Pythagoras (fl. late 500s B.C.) and the Pythagoreans considered number to be the foundation of everything. The monad (or the one) is the principle of all things, maintained Pythagoras. From the monad arises the dyad, or two; from one and two come all numbers; from numbers arise geometric points; from points, lines; from lines, plane figures; from plane figures come solids; from solid figures arise sensible bodies, "the elements of these being four, fire, water, earth, and air."[8] In Plato's *Timaeus,* a Pythagorean-like doctrine is offered wherein these four elements actually have geometric forms—fire being a tetrahedron, air an octahedron, water an icosahedron, and earth a cube—and these elements change into one another according to definite mathematical ratios.[9]

Also, too, there were the early atomists, Leucippus, Democritus and, later, Epicurus, contending that the universe comprises an infinite number of indivisible units in a void—atoms too small to be detected by the senses, dif-

---

[8]Diogenes Laertius *Lives of the Eminent Philosophers* 8.25.

[9]We are ignoring, in this quick discussion of Plato, the extraordinary complexity of his philosophic arguments, as well as questions regarding the actual "spirit" in which the arguments in the *Timaeus* should be taken (given that the arguments are put into the mouth not of Socrates but of Timaeus and are expressed with reservations).

fering in size and shape, and possessing no other qualities except those of impenetrability and solidity. They were apparently wrong about atoms being impenetrable, absolutely solid and infinite in number, and they merely assumed rather than tried to explain the origin of motion; but their description turned out to be in some respects surprisingly close to the mark. The irony is that it was far too untethered from what was observable at the time to inform scientific investigation.

Much as later physicists needed Newton's formulations concerning motion on the way to Einstein's theory of relativity, early explorers of the chemical order needed a model a good deal more down to earth on their way to modern chemistry—something in closer contact with everyday experiences—if they were to move forward in their exploration.

Aristotle (384-322 B.C.) provided a model that turned out to be both misleading and leading, erroneous in key aspects, but also, surprisingly fruitful. Aristotle's core concept of form, moreover, anticipated by more than two thousand years an emerging thread in twentieth-century science. For him, Plato's most famous pupil, the four elements were taken up into a sophisticated philosophical framework in which the ultimate cause of structure and organization was form (in Greek, *eidos*). In physical things, forms were united to matter (the matter being relative to each kind of form), and these forms gave to indefinite matter its definite shape, powers and properties. Earth, water, air and fire came (recalling Anaximander) from the mixture of the more elemental properties hot, dry, moist and cold.[10] Aristotle's account, with slight variations, guided the intellectual understanding of chemistry for nearly two millennia.

We have, then, prior to the birth of Christ, two modes of activity, both of which were essential to the penetration of the order of physical reality and, hence, without which the order of the periodic table of elements would never have been discovered. On the one hand, we have those craftsmen who accumulated a vast body of practical knowledge in their unceasing effort to produce highly impractical, beautiful objects. These craftsmen were well versed in the *how to*, but had insufficient appreciation of the *what* and the *why*. They immersed themselves in the reality of the chemical order without inquiring into the ultimate nature of that reality and, hence, into *what* things really were and *why* their technical artistry actually worked.

---

[10]As with Plato, this quick account in no way does service to the actual complexity of Aristotle's arguments.

These questions awaited the philosophers who came later. In their attempts there was something grand; they were plunging beneath the welter of craftsmen's tricks in search of something essential, permanent. And yet, however intellectually inviting and satisfying, their explanations, as we have seen, were woefully insufficient in explaining the actual complexities encountered every day by the metalworkers, dyers of cloth and pottery makers. The philosophers were rightly eager to elucidate the nature of reality, to find what was truly elemental, but the temptation was always to confuse the world of human thought with the actual complexities of the real world, to be content with a tidy intellectual simplicity which only a long apprenticeship in the untidy complexities of a laboratory could cure.

To advance our understanding of what was truly elemental, humanity, then, needed a kind of union of the artisan, immersed in the actual complexities of reality, and the philosopher, immersed in the attempt to discover the underlying order of these complexities. In the centuries around the birth of Christ, this union was achieved, however imperfectly, in the person of the alchemist, whose technical search for the way to turn base metals into gold was animated by a philosophic search to discover the innermost secrets of nature's order. Alchemy has been all too quickly made the object of scorn, a kind of byword for any fantastic and fruitless pseudoscientific endeavor. That is a half-truth. Alchemy was indeed fantastic insofar as its activities were shrouded in an almost impenetrable mysticism of strange terms and procedures.[11] But it was far from fruitless, even though its adherents never produced what they so passionately sought.

Alchemy operated under that quasiscientific, quasimagical notion that there was a kind of unity of matter and, more importantly, that matter could therefore be transmuted into its most perfect form, gold—if only someone could discover the philosopher's stone that, as a "medicine of the metals," would cure the base metals (like lead or tin) of their baseness. Interestingly, the most common belief concerning the essentials of this transformation involved the use of two elements, sulfur and mercury. Sulfur was to act as the

---

[11]On an alchemist's shelf you might find vitriol of Cyprus, white vitriol, aqua regia, arsenic, saltpeter, alum of Yemen, burning water, brimstone, sal ammoniac, sophic salt and quicksilver. For the most part, these are names of compounds (e.g., the two vitriols are, respectively, copper sulfate and zinc sulfate). We also note the existence of the actual chemical elements arsenic (As); brimstone, which we call by the less poetic name sulfur (S); and quicksilver, which we call mercury (Hg). We must realize, however, that alchemists did not consider these to be elemental. As is the case innumerable times throughout human history, having something important right under one's nose doesn't mean that its actual importance is recognized.

principle of fire (hot and dry) and mercury of water (cold and moist), and their felicitous combination—using the four ancient Aristotelian principles of hot, dry, cold and moist—would produce gold.

In their vain search for the philosopher's stone, alchemists amassed an enormous body of knowledge about chemical procedures, substances and laboratory apparatus. They perfected the essential chemical arts of distillation and sublimation and, in doing so, created the laboratory apparatus for these arts, such as the alembic, the teardrop-shaped glass container with the long neck. As they were looking for the precise way of creating gold, they were very precise in making and recording combinations of chemicals and procedures. They distinguished a host of chemical compounds by their distinctive properties. And they tamed fire in an effort to control it more precisely under laboratory conditions. On this foundation, modern chemistry was built—a sign of this being that it is a real matter of controversy among historians of science as to when alchemy became chemistry.

In accurately assessing the alchemists' labors, we must then ask ourselves, would anything less than the alchemists' passion have sustained their efforts in the long, sweat- and smoke-filled laboratories, day after day, year after year over the millennium-and-a-half reign of alchemy? The great German chemist Justus von Liebig (1803-1873) thought not:

> The most lively imagination is not capable of devising a thought which could have acted more powerfully and constantly on the minds and faculties of men, than that very idea of the Philosopher's Stone. Without this idea, chemistry would not now stand in its present perfection. . . . In order to know that the Philosopher's Stone did not really exist, it was indispensable that every substance accessible . . . should be observed and examined. . . . But it is precisely in this that we perceive the almost miraculous influence of the idea.[12]

Without the grand failure of alchemy, there would have been no chemistry. The search for the philosopher's stone was one of the most fruitful of fruitless adventures. There was no necessity that humanity ever progress beyond the division betweens artisans with immense practical knowledge for the production of beauty on the one hand and, on the other, philosophers with their abstract focus on elaborating nature's underlying unity. Indeed, for most civilizations this duality was never overcome. The union of the ar-

---

[12]Quoted in John Read, *From Alchemy to Chemistry* (1961; New York: Dover, 1995), p. 29. Read's book is an excellent, concise overview of alchemy and its relationship to modern chemistry.

tisan and the philosopher in the alchemist was a first and crucial phase in overcoming this duality.

The spirit animating this union, if we purify it and examine its true nature, was the very scientific, very human belief that there is a magnificent underlying order of nature waiting to be discovered—a secret like the philosopher's stone—if only we patiently search through the actual particulars of nature. That is, scientific exploration assumes that there exists an underlying order of the world that is intelligible even when it is yet undiscovered, a secret code ciphered into the natures of things themselves, a knowable order rather than mere gibberish. This much, at least, the alchemists understood. That having been said, the absolutely necessary mediating phase of alchemy itself had to be overcome. For this to happen, the truth itself about nature, and about the elements, had to become even more precious than gold.

A next step was taken during the Renaissance in the turn from the attempt to produce gold toward efforts to create effective medicine for the human body. This was a transformation of alchemy's goals, even while it bore a deep continuity. The philosopher's stone was also known as the *elixir vitae,* the elixir of life, because it was considered not only the perfect medicine to cure base metals (like lead) of their baseness, but also the perfect medicine to cure our own bodily maladies (and perhaps our mortality). The name associated with this transformation from alchemy to iatrochemistry (chemistry as applied to health, from the Greek word for healer, *iatros*) was the infamous Philippus Theophrastus Aureolus Bombast von Hohenheim (1493-1541), or Paracelsus, as he dubbed himself. Paracelsus believed in the famous four elements as enunciated by Aristotle and developed by the alchemists, but he asserted that they appeared in all bodies (inorganic and organic) as three principles: salt, sulfur and mercury, the so-called *tria prima.* Thus, the focus of iatrochemistry was not the alchemical making of gold from sulfur and mercury, but the chemical curing of the human body's chemical imbalances.

Paracelsus waxed eloquently on the curative powers of antimony, unbeknownst to him another element (Sb) on the periodic table, an element known in compound form to the ancients (as astibnite, or antimony sulfide) and used to darken women's eyebrows! We now understand that his notion of antimony as an elixir was dangerously wide of the mark. But his theory served to move the protochemists of the time beyond the alchemical search for the philosopher's stone, which by then had become merely ornate, affected, repetitious and theoretically impotent. For while alchemy had advanced knowledge of many laboratory procedures and had distinguished a multitude of different compounds, cataloguing their properties and various effects, its single goal of producing gold gave it a focus ultimately too constrained, too myopic; and therefore it ended in running its adherents through the same intellectual ruts.

Paracelsus's great service was that, in directing attention toward a new problem, alchemists were bumped out of their ruts and liberated to perform new kinds of experiments. No longer were they after the answer to a single problem. Ultimately Paracelsus's four humors became one big quagmire, but an important step had been taken forward that would not be taken back. In arguing that various human maladies could be cured by as-yet-undiscovered chemical remedies (rather than the herbal, organic remedies handed down from Greco-Roman medicine), Paracelsus opened up many new paths of experimental inquiry.

It was Robert Boyle (1627-1692) who, we might say, chose the fateful path, the trail that would lead chemistry to its proper object. Boyle's step was not the grand end of the quest. As with every great quest, the object lay at such a distance that its true contours were, as yet, only dimly perceived. Boyle, however, took the decisive first step down a path that was just off the main highway and little more than a small break in the woods and yet one through which a distant shaft of light was barely visible. Boyle's contribution was simply to insist that the study of chemistry was worth doing for its own sake, health and wealth being dross by comparison.

It is worth noting also that this turn to the study of nature for its own sake was historically facilitated by a theological position called voluntarism, which turned away from rationally deductive systems to the study of nature itself as it revealed God's will. Voluntarism itself was a reaction to the overly zealous Aristotelianism (technically called Latin Averroism, after the Islamic philosopher Averroës) that sought to constrict God's creative power according to the rational confines of Aristotle's physics. The fourteenth-century theologian and philosopher William of Ockham and his followers denied that God was restricted by any such necessity in creating the world but, instead, that he could create it any way he willed (Latin *voluntas,* meaning "will" or "wish"). Because there was no necessity in what God happened to will, no human, rationally deductive philosophy could reveal the order of nature. Instead, we must turn to nature itself to uncover the particular way God happened to create. While this in turn helped push scientists to stress actual experiment—a healthy development—voluntarism, in its adamant rejection of Aristotle, tended to lead to nominalism, the belief that there are no universals in nature such as "cat," "dog" and even "human being," and hence that universal names (Latin, *nominees*) were mere human constructions. Nominalism, in turn, undergirded much of modern scientific reductionism.

Whatever the larger historical antecedents of Boyle's turn, we certainly have in Boyle a more profound union of the philosopher and the artisan than occurred in either alchemy or in the earliest medical chemistry. Without this union, there would be no science of chemistry.

This is not mere hyperbole. To advance, the early chemists needed even greater liberty than the medical motivations Paracelsus provided. The rich complexity of nature so far exceeds any narrow use the early chemists might have hoped to wring from it for health that without a grail as broad and deep as truth itself, they could never have plumbed the chemical order. The experiments instrumental in revealing the elements and their relationships were, as far as the early modern chemists knew, quite without any visible practical benefit. The animating spirit of this experimentation was not utility, not the desire for gold or health; even less was it the desire for sex or self-preservation. The immediate cause of our deep insight into nature's chemical order, and hence the science of chemistry, was the most human and singular desire to know the truth about the chemical elements.

Boyle's second major contribution was related to his first. He set down and disseminated a sound definition of *element* and, hence, defined what chemistry, as a science, should search for.[13] As he stated in the sixth part of *The Sceptical Chymist,*

> I now mean by elements, as those chymists that speak plainest do by their principles, certain primitive and simple, or perfectly unmingled bodies; which not being made of any other bodies, or of one another, are the ingredients of which all those called perfectly mixt bodies are immediately compounded, and into which they are ultimately resolved.[14]

At this time it was unclear what was elemental, but this plain-speaking definition brought to the fore the all-important distinction between chemical elements defined as simple and unmixed, and chemical compounds defined

---

[13]As we shall see, Boyle was an atomist and considered atoms to be the true element; he considered what we would now consider elemental (e.g., gold) as being secondary compounds that he called "primary concretions."

[14]Quoted in William Brock, *The Norton History of Chemistry* (New York: W. W. Norton, 1992), p. 68.

as mixtures or unions of elements. The job of chemistry became figuring out just what was really elemental—the exact task undertaken by the ancient philosophers. But unlike the ancient philosophers, Boyle asserted that to uncover what is truly elemental, we could not be satisfied with exalted abstractions; we must throw ourselves into the sweaty, laborious trials of the artisans and fight our way, inch by inch, through the actual complexities of chemical nature.

As a token of his intentions to replace philosophic abstraction with well-grounded observation, Boyle dismantled the four-element theory that stretched back to Anaximander, heedless that Aristotle's almost hallowed authority stood behind it. Again, according to Aristotle, earth, air, fire and water were the primary elements and everything else was a mixture of these. As proof, it became customary to demonstrate the existence of the four elements in things by putting a green stick in a fire. The *fire* trapped within was released as it burned, the smoke that curled up was escaping *air*, the liquid that oozed out of the ends of the stick was *water*, and the ash left over was *earth*. Hence, wood is a mixture of fire, air, water, and earth! To unhinge this facile demonstration, Boyle carefully analyzed the alleged elements and concluded that "as for the greene sticke, the fire dos not separate it into elements, but into mixed bodies." The fire is "but the sulphurous part" of the stick "kindled"; the "water boyling out at the ends" is not "elementary water" but a mixture, a sign of which is that the juices of plants are used "by physitians" to cure "several distempers" against which water itself is ineffective; the "smoake" is a "very mixt body" that when distilled yields "an oile," "abounds in salt," tastes bitter and, furthermore, makes one's eyes water "which the smoake [i.e., steam] of common water will not doe."[15]

Equally telling against the four-element theory, "out of some bodies, four elements cannot be extracted, as Gold, out of which not so much as any one of them hath been hitherto [produced]."[16] In sum, too many things are extracted from the "greene sticke"—and they are compounds (mixed bodies), not elements (simple bodies)—and nothing at all is extracted from some materials, like gold.[17] Such is the triumph of experiment over theory, and such too is the triumph of the order of nature over human speculation. As convincing and self-consistent as a theory or some aspect of a theory of nature

---

[15]Quoted in ibid., p. 57.
[16]Quoted in ibid.
[17]Interestingly, even though Boyle did not draw this conclusion, gold satisfied his definition of an element as "primitive and simple, or perfectly unmingled."

may be, as helpful as it may be as a stepping stone for advancing our understanding of nature, it is ultimately the order in nature that determines the accuracy of the theory.

Boyle helped consign Aristotle's four-element theory to the history of science by pointing us away from the theory to the actual intricacies of nature before our eyes. It's curious, then, that there was one part of Boyle's view of nature that wasn't really accessible to experimentation (at least not for several centuries). His theory, which he used as a foil against Aristotelianism, was the Epicurean-Democritan version of atomism made known to him largely through the works of Pierre Gassendi (1592-1655). As noted above, what was truly elemental on this view was bulk homogenous matter differing only in size, shape and position. Boyle argued that when these different-sized and -shaped atoms came into particular combinations, they made "primary concretions," which corresponded only roughly with what we now consider a chemical element. Thus, Boyle partly disagreed with the definition given by "those chymists that speak plainest." On his view it was the homogenous atom—and not any visible substance like gold, silver or lead—that was elemental. The qualities of gold, silver and lead were merely the result of the way homogenous atoms of differing size and shape fit together. Gold, therefore, was not an element for Boyle. In an important sense, Boyle was right. His description of "atoms" comes very close to the modern understanding of protons, neutrons and electrons. While we now know that not even the proton, neutron and electron are themselves fundamental, and although they differ from each other in more than just size, shape and position, their existing together in particular combinations does result in the particular elements (gold, silver, etc.).

But Boyle's speculations on this score were, at best, weakly grounded empirically. Scientists then lacked the instrumentation to gain purchase on Boyle's model. Thus, chemists partook of Boyle's own skepticism (made famous in his *Sceptical Chymist*) about theories held against or beyond the available evidence, and they generally ignored his theoretical atomism, preferring to keep to the visible, tangible order amenable to laboratory analysis. There was too large a gap between the invisible order alleged by his atomism, and the rich and variegated order of chemistry as a laboratory science. So it was that, contrary to the claims of many popular histories of science, chemists remained skeptical of such atomism into the mid 1800s.[18] Of course, they had good reasons to do so. The actual chemical order in front of them and the multitude of complex changes and qualities produced and examined in their daily work were not demonstrably reducible to the mere size, position and shape of some invisible, homogenous atomic entity.

Boyle died near the very end of the seventeenth century. Up to that point, the only known elements were gold (Au), silver (Ag), mercury (Hg), copper (Cu), lead (Pb), tin (Sn), iron (Fe), sulfur (S), carbon (C), arsenic (As), antimony (Sb), bismuth (Bi), zinc (Zn), and phosphorus (P). To be more accurate, while they were known as substances, they weren't known as elements. As the list makes clear, they were, for the most part, metals, and they were isolated and identified not out of any philosophical-chemical prowess but because (fortunately) they occurred in large enough agglomerations in ores near the surface of the Earth. Even when attempts were made to determine whether they were elemental—that is, simple substances rather than compounds—the problem lay in how to identify whether, in some chemical reaction, the thing you had to begin with or the thing that you got as a result was truly elemental. The eighteenth century was the watershed in identifying what was truly elemental, and it took earth, air, fire and especially water to do it.

It should make us wonder that water is so ubiquitous, so life giving and yet so simple. Water's presence everywhere on our planet and its intimate involvement with our existence make it an object of our especial focus, as is clear from its exalted position, for over two millennia, as one of the four elements. Of course, since water is a most excellent solvent, it is found in every laboratory, from those of the present day back to those of the earliest alchemists. What is quite strange, quite unexpected, is that while water is not itself an element, it is one of the simplest of compounds—about the shortest step possible from elements to a compound. That simplicity made it relatively easy to take it apart, to recognize the two parts as elemental and to put it back together again—an enormous intellectual leap. Imagine how difficult such analysis would be if the liquid of life was not made of only two elements (hydrogen and oxygen) in a very simple ratio of 2:1 (H_2O), but instead was on the level of complexity of, say, glycerol ($CH_2OHCHOHCH_2OH$). It would be like trying to learn to read using T. S. Eliot's poem "Sweeney Among the Nightingales" as a primer. As compared to such complex compounds, water was a relatively easy "primer" to read.

---

[18]For a nuanced discussion about skepticism toward atomism in the history of chemistry, see William Brock, *Norton History,* pp. 165-72.

But we must stress the qualifier *relatively*. As we shall see, it was quite a struggle, and that struggle will make clear the importance of nature's condescension as a kind of tutor in providing us this simple compound as a key to learning to read its chemical order.

The discovery that water was not itself an element was made possible through the use of fire, air, and earth (in this case, metals). It begins with a very common laboratory operation, known even by the alchemists, wherein the burning of some metal in the open air caused it to calcinate (i.e., to turn white). For example, when zinc was burned, it turned whitish (hence the burned zinc was called calx of zinc, *calx* being Latin for lime or chalk). Amazingly, when the metal calx was reheated, this time with charcoal, it would magically return to its pristine condition. A lovely trick! Needless to say, the alchemists got a lot of mileage out of this apparent magic, if not in advancing their efforts to produce gold, at least in convincing others of their mystical powers.

But what was actually going on? No headway on the matter could be made until someone paid attention not only to earth (the metal) and fire, but also to air. It was Joseph Priestley (1733-1804) who did so. Priestley took metal calx (in this instance, calx of mercury), put it under a glass container from which a tube ran to a pneumatic trough, trapping any gas released.[19] Using a magnifying glass, Priestley then concentrated the burning rays of the sun on the calx of mercury. The calx of mercury turned back into pure mercury, and Priestley trapped the gas. The gas was, as we now know, oxygen; and calx of mercury is actually mercuric oxide (HgO). The oxygen was added when the mercury was burned in the open air; it was removed when the mercury was burned under the glass.

We might surmise from this that Priestley had discovered oxygen; and indeed, he is almost invariably given the credit by textbooks and histories alike for having done so. The truth is more complicated. Priestley didn't discover oxygen because he thought that the gas he had trapped was actually *dephlogisticated* air. In fact, he went to his deathbed vehemently denying that oxygen even existed. What is this theoretical substance, dephlogisticated air? Priestley was a diehard adherent of the phlogiston theory formulated by Johann Becher (1635-1682) and popularized by Georg Ernst Stahl

---

[19]Pneumatic trough is a fancy name for an upside-down glass filled with liquid that sits in a basin filled with the same liquid, into which a tube runs so that any gas running into the tube will be trapped in the upside-down glass (displacing the liquid).

(1660-1734). According to this theory, things that burned had phlogiston in them (*phlogiston* is the Greek word for "burned" or "set on fire"), and the phlogiston was released in burning.

If, for example, you placed a lit candle under a glass globe, as Priestley and others had done so many times, eventually (so they maintained) all the phlogiston would be released from the candle and saturate the globe. When the air in the globe could hold no more phlogiston, the candle went out. In regard to the burning of metal, since the pure metal burned red-hot, it must (supposedly) have phlogiston in it. As the phlogiston was steadily released, the metal degraded, like a burnt log. Thus, the metal calx was dephlogisticated. Burning the calx again, especially with charcoal (which, everyone knew, must be rich in phlogiston!), caused the phlogiston to rush back into the metal and restore it to its natural, pristine condition. According to phlogiston theory, pure metal, then, was not really pure; that is, it was not really an element but a kind of compound. Zinc, for example, was really zinc plus phlogiston.

Returning to the above experiment with mercury, Priestley reasoned that since the calx of mercury was being restored by the sun's rays to mercury, its phlogiston was returning, and the air itself (trapped in the pneumatic trough) must have been dephlogisticated. He had a further proof of this. If you place a candle under a glass globe filled with dephlogisticated air (which was, as we now know, oxygen), the candle burned with "a remarkably vigorous flame."[20] That seemed to explain why the candle burned so brightly. Since it was dephlogisticated air, argued Priestley, the phlogiston rushed out of the candle and into the phlogiston void with unusual vigor.

But French chemist Antoine Lavoisier (1743-1794) came to believe something quite different. He contended that phlogiston wasn't being released from metal as it was burned. Indeed, he asserted that phlogiston did not exist. Instead, the metal was taking in oxygen. Calx of mercury, he argued, was really mercuric oxide. Further, the mythical phlogiston wasn't returning when calx of mercury was again burned; rather, oxygen was escaping from the mercuric oxide, leaving plain mercury, a bona fide element. Nor did the candle under the glass burn out because the glass was saturated with phlogiston. No, it was snuffed out because the flame had used up all the oxygen

---

[20]Quoted in J. R. Partington, *A Short History of Chemistry*, 3rd ed., revised and enlarged (New York: Dover Publications, Inc., 1989), p. 118. The whole section on Priestley in the work is quite clear and insightful, especially in regard to his adherence to the theory of phlogiston.

in the glass. By the same token, the candle burned more vigorously in this strange, trapped air not because the air was dephlogisticated and the phlogiston could gush forth freely from the candle, but because the strange air was pure oxygen, the very food of flame.

But how to demonstrate that he was right and Priestley was wrong? Both descriptions appeared to account for the observed phenomena equally well. Lavoisier happened to meet Priestley in 1774 in Paris. At a dinner, he listened carefully to Priestley's account of dephlogistication and recognized that something did not add up. To be exact, as many others had known even before Priestley, the metal calx always weighed *more* than the pristine metal, a sign that something was being added, not taken away when the metal was burned. Phlogistians avoided this embarrassment by arguing that phlogiston had negative weight, and so when phlogiston was in the metal it weighed less than the dephlogisticated metal calx! Lavoisier found a way around the impasse. He did much the same experiment as Priestley except he left a little air (50 cubic inches) in the upside-down glass (called a bell jar) sitting in the pneumatic trough filled with liquid. As with Priestley, the bell jar was used for capturing air through a tube that ran between the bell jar and another glass container (called a retort) in which the chemist heated various substances. Using this set-up, Lavoisier devised a clever way of testing Priestley's theory.

As he suspected, he found that when pure mercury was burned in the retort, the 50 cubic inches of air left in the bell jar shrank down to 42 cubic inches. In other words, air had been sucked out of the bell jar, back through the tube and into the retort where the mercury was being burned—proof that, upon burning, air was going into the mercury rather than phlogiston going out. Thus, the burned mercury, calx of mercury, was actually mercury plus air. That is why metal calx weighed more than pure metal, Lavoisier reasoned. What kind of air was going in? Lavoisier now did the experiment in the reverse, burning calx of mercury and trapping the gas given off. To this gas, he gave the name oxygen.[21] And so it was that Lavoisier was able to isolate a true element—the hitherto hidden element, oxygen, one of the two elements needed to make water.

Who discovered the other "half"? Generally, the discovery of hydrogen is credited to Henry Cavendish (1731-1810), but we have the same kind of

---

[21] *Oxygen* means "acid bearer" in Greek. Lavoisier incorrectly thought all acids contained oxygen.

story to tell here. Cavendish did indeed isolate hydrogen, which, since it burned, he called "inflammable air." But because it burned, he thought that he had isolated the famed phlogiston. He then did something even more amazing but misunderstood (especially by him). He added fire to the two "airs" and made water; that is, he put inflammable air (hydrogen) and dephlogisticated air (oxygen) into a glass globe and sparked them. The result? Water! How could this be? Attempting to fit it into the reigning phlogiston theory, he reasoned, albeit cautiously,

> that dephlogisticated air is in reality nothing but dephlogisticated water, or water deprived of its phlogiston; or, in other words, that water consists of dephlogisticated air united to phlogiston; and that inflammable air is either pure phlogiston . . . or else water united to phlogiston.[22]

Therefore, the amazing result could be readily summarized in an equation: inflammable air [water + phlogiston] + dephlogisticated air [water − phlogiston] = water [dephlogisticated air united to phlogiston]. Simple (but highly dubious) math!

This explanation did not content Lavoisier, of course. He had become convinced that phlogiston simply didn't exist. He thus isolated Cavendish's inflammable air, sparked it with oxygen and made his own water. He then "decomposed" water (in the form of steam) into "inflammable air" and oxygen. Since he knew that oxygen was elemental and that phlogiston didn't exist, he reasoned that inflammable air must actually be an element. This element he called hydrogen, or "water bearer." So it was that water, with the help of earth and fire, was finally proven not to be elemental, but to be made of air (or to be more exact, two gases). Further, Lavoisier had shown that the metals themselves were not mixtures but pure elements according to the definition of element that Boyle had rejected. Additionally, Lavoisier showed that air contains more than just oxygen by demonstrating that the air left over after all the oxygen is burned out, Priestley's "phlogisticated air," was a distinct element as well, which he named azote (but was later called nitrogen). Hence, Lavoisier eliminated not only water as a simple element, but air as well.

In the midst of his labors combating the confusions of phlogiston theory, Lavoisier uttered a famous complaint about the phlogistians that is well worth quoting, as it parallels complaints made by reputable scientists against natural

---

[22]Quoted in Partington, pp. 139-40.

selection as a do-anything, do-everything mechanism for Darwinism.[23]

> Chemists have made phlogiston a vague principle, which is not strictly defined and which consequently fits all the explanations demanded of it. Sometimes it has weight, sometimes it has not; sometimes it is free fire, sometimes it is fire combined with an earth; sometimes it passes through the pores of vessels, sometimes they are impenetrable to it. It explains at once causticity and non-causticity, transparency and opacity, color and the absence of colors. It is a veritable Proteus that changes its form every instant![24]

Lavoisier was a true revolutionary, cutting his way through both ancient and modern errors and leading us to the elements as elements. Unlike Boyle, Lavoisier really did believe in the same definition of element as "those chymists that speak plainest." In his formulation, an element was "the last point which analysis is capable of reaching." As with water, proving something to be the "last point" is tricky since, as Lavoisier cautioned, substances may appear to us to be simple only because "we have not hitherto discovered the means of separating them." If we look at Lavoisier's list of elements—and it is truly chemistry's first real list of elements *as* elements—we find him to be amazingly accurate, correctly identifying as elemental oxygen, azote (nitrogen), hydrogen, sulfur, phosphorous, carbon, antimony, arsenic, bismuth, cobalt, copper, tin, iron, lead, manganese, mercury, molybdenum, nickel, platinum, silver, gold, tin, tungsten and zinc.

The list, not surprisingly, was less than perfect. You would also find, listed as elements, what we now know to be compounds, such as lime and magnesia. You would even find light and heat (or caloric). Yet it was a tremendous leap forward. Read any chemistry text or treatise prior to Lavoisier's revolution, and we find an impenetrable morass of strange terms, procedures and beliefs. Read Lavoisier's own works, and suddenly we find ourselves on familiar ground. Lavoisier and his framework for understanding elements opened the floodgate. Chemists quickly discovered element after element, with over 75 percent of the elements on the periodic table identi-

---

[23]See, e.g., Philip S. Skell, "Why Do We Invoke Darwin," *The Scientist* 19, no. 4 (Aug. 29, 2005): "Darwinian explanations for such things are often too supple: Natural selection makes humans self-centered and aggressive—except when it makes them altruistic and peaceable. Or natural selection produces virile men who eagerly spread their seed—except when it prefers men who are faithful protectors and providers. When an explanation is so supple that it can explain any behavior, it is difficult to test it experimentally, much less use it as a catalyst for scientific discovery."
[24]Quoted in Brock, *Norton History*, pp. 111-12.

fied within 150 years of Lavoisier's death.

The rush of discovery was borne along by water and galvanized by electricity, a close "relative" of fire. Prior to the nineteenth century, there was no electricity in laboratory use except static electricity. What was needed to further research was a continuous, readily available current. Italian physicist Alessandro Volta (1745-1827) found that when dissimilar kinds of metals touched his tongue and were brought into contact, a bitter taste resulted, and when he similarly touched his eye, the metals somehow created a sensation of light. He then tried piling various metals on top of one another (e.g., zinc and silver disks), alternating them with blotting paper soaked in brine. He found that they generated a continuous electric current, albeit a weak one. Volta reported this, and soon scientists throughout Europe were creating larger and larger voltaic piles, for the larger the piles, the stronger the continuous electrical current. In England, avid experimenters Anthony Carlisle (1768-1840) and William Nicholson (1753-1815) wondered what effect this current would have on water. The effect was . . . bubbles—in particular, the bubbles of two gases, the recently discovered hydrogen and oxygen. Electricity somehow shocked water into its constitutive elements. Once it was understood that electricity had this effect, electrolysis (decomposition through the use of an electric current) became a key to discovering a whole string of hidden elements, the most famous and successful chemist in this regard being the great Humphrey Davy (1778-1829), who, in a matter of two years (1807-1808) discovered potassium (K), sodium (Na), barium (Ba), strontium (Sr), calcium (Ca) and magnesium (Mg).

Even more important, however, was another insight afforded by electrolysis: that elements must somehow be electrical in nature (making them quite lively and hence something quite different from Boyle's homogenous and inert blobs of matter). To Davy, if electricity broke apart the most stubborn compounds, then somehow their union must be electrical—especially since the hydrogen always collected at what Volta had designated the negative pole and oxygen at the positive pole. Although it would take a century to uncover the causes, this was an early clue to the underlying nature of the elements, revealing as it did the effects of the positively charged proton and the negatively charged electron.

While electrolysis led to the discovery of many new elements, having identified a list of elements is not the same as having them laid out *in order* on the periodic table as part of a rich understanding of what made them tick. More clues would have to be provided by nature before the order of the elements on the table could be unveiled. Fortunately, scientists refused to believe that nature contained merely a large, haphazard pile of elements. They had always found nature too ingeniously contrived for that and were therefore justly suspicious that a meaningful order lay hidden.

This suspicion of order is rooted in our very human love of patterns and predates the scientific breakthroughs that confirmed it. We cannot believe that nature does things randomly, a sign of which (ironically) is that even dogma-bound materialists, who insist that the universe has no designer, are convinced that nature is governed by laws—and not just any laws, but intelligible, even elegant laws. Just as ancient geometers assumed the existence of an elegantly contrived geometric cosmos, so also modern chemists suspected that the elements were ordered in some elegantly contrived and harmonious way. As is almost always the case, the underlying order was far more elegant and harmonious than even the most conspiracy-minded chemists could imagine. They suspected a beautiful melody. They discovered a symphony.

We are aware today that atoms are a little like our solar system in that a much larger nucleus (made of one or more protons and neutrons) is orbited by one or more far smaller particles called electrons. Refer again to the periodic table provided at the beginning of the chapter (see p. 112). The elements line up horizontally on the table according to the number of protons each has in its nucleus, and this number is called the "atomic number." Hence, hydrogen (H) has one proton and is atomic number 1; helium (He) two protons, atomic number 2; lithium (Li) three protons, atomic number 3; beryllium (Be), four, atomic number 4, and so on as one moves from left to right across the chart and then back again to the left side to start the next horizontal row. (The horizontal rows are called periods, hence the *periodic* table of elements.)

Also, we now understand that in each atom there is an electron for every proton and that these electrons orbit the nucleus in precisely defined energy levels, or shells. The elements line up "vertically" according to the number of electrons orbiting the nucleus in the outermost energy level or shell of the atom. Thus, all the elements in the first vertical row (or "group," as the vertical rows are called) have one electron in the outermost level; all the elements in Group II have two electrons in the outermost level. Leaping over to Group III (sometimes denoted Group IIIA to distinguish it from Group IIIB of the transition elements)—the vertical row with boron (B) at the top— each element has three electrons in its outermost level, and so on, as we move further to the right on the periodic table (leaving aside the rather more complex transition elements, the lower-slung "bridge" of elements that fits in the gap defined by the leap from calcium, Ca, to gallium, Ga).[25] Lined up thus, the table would illuminate a variety of telltale patterns and relationships among the different kinds of atoms.

But of course, the scientists who actually cracked the code of the table's order a century and a half ago were not isolating individual atoms—however they may have conceived them—pulling them apart with Lilliputian tweezers and stacking up protons and electrons for easy reckoning. There is such an enormous gap between the everyday level of substances dealt with by chemists and the subatomic order, that if we had first to understand the subatomic world of protons and electrons to set the elements in order on the periodic table, then there never would have been a periodic table. Fortunately, nature is an accommodating tutor. There are layers of order mediating the subatomic world to the everyday world.

By the early 1800s there were enough elements known *as* elements that scientists could begin to look for patterns in the "pile" of elements before them. Atomic weight was measured rather crudely at this point, according to a scale relative to a chosen element (e.g., hydrogen) as measured by equal volumes or parts of elements as they were combined in the laboratory. Interestingly, even this crude method led to the discovery of real, underlying patterns. One of the first to note patterns among the known elements was Johann Döbereiner (1780-1849), who picked up on what he called "triads." Triads were groups of three elements wherein the atomic weight of the middle element was (approximately) the mean between two other elements. For example, Döbereiner noted that the weight of strontium (Sr) was a mean between the weights of calcium (Ca) and barium (Ba). He found other triads as well: lithium (Li), sodium (Na) and potassium (K); sulfur (S), selenium (Se) and tellurium (Te); and chlorine (Cl), bromine (Br) and iodine (I). If you look on the periodic table—the table that, of course, had yet to be discovered—you can see that these triads are lined up vertically in different groups (fig. 5.2).

This was an amazing discovery, especially since Döbereiner (through no

---

[25]The transition elements (or transition metals, as they are often called) form a "bridge" between those elements with two electrons in the outer shell and those with three. A glance at a periodic table that displays the electron shell structure reveals that these transition elements build electrons not in the outer shell but in the inner shells. As a result, their properties vary only slightly, and hence form a very gradual transition of qualities from Group 2 to Group 3.

1 H																	2 He
3 Li	4 Be											5 B	6 C	7 N	8 O	9 F	10 Ne
11 Na	12 Mg											13 Al	14 Si	15 P	16 S	17 Cl	18 Ar
19 K	20 Ca	21 Sc	22 Ti	23 V	24 Cr	25 Mn	26 Fe	27 Co	28 Ni	29 Cu	30 Zn	31 Ga	32 Ge	33 As	34 Se	35 Br	36 Kr
37 Rb	38 Sr	39 Y	40 Zr	41 Nb	42 Mo	43 Tc	44 Ru	45 Rh	46 Pd	47 Ag	48 Cd	49 In	50 Sn	51 Sb	52 Te	53 I	54 Xe
55 Cs	56 Ba	57 * La	72 Hf	73 Ta	74 W	75 Re	76 Os	77 Ir	78 Pt	79 Au	80 Hg	81 Tl	82 Pb	83 Bi	84 Po	85 At	86 Rn
87 Fr	88 Ra	89 ** Ac	104 Rf	105 Db	106 Sg	107 Bh	108 Hs	109 Mt	110 Ds	111 Uuu	112 Uub						

*	58 Ce	59 Pr	60 Nd	61 Pm	62 Sm	63 Eu	64 Gd	65 Tb	66 Dy	67 Ho	68 Er	69 Tm	70 Yb	71 Lu
**	90 Th	91 Pa	92 U	93 Np	94 Pu	95 Am	96 Cm	97 Bk	98 Cf	99 Es	100 Fm	101 Mb	102 No	103 Lr

Figure 5.2.

fault of his own) was using incorrect weights. Yet nature and previous technological advances had given him the tools to get close enough for him to
see a pattern, even though he couldn't possibly know the ultimate causes of
the pattern. Those would have to wait about another century until further
aspects of the order had been revealed.

About the mid-1800s, further attempts were made at deciphering the underlying order. This time, using estimated atomic weights that were far more
accurate, scientists searched for an elegant mathematical relationship between the atomic *weights* of elements and the *properties* of the elements.
The underlying belief was that, somehow, elements with similar chemical
properties must be related to each other by some kind of mathematical pattern via weight. Again, we find that central to scientific discovery is the suspicion of order.

Jean Baptiste André Dumas (1800-1884) noticed that elements that had
the same chemical properties and increasing atomic weights could be related by simple mathematical patterns. Here were the atomic weights available to him:

nitrogen (N) = atomic weight 14
phosphorous (P) = atomic weight 31 [14 + 17]
arsenic (As) = atomic weight 75 [14 + 17 + (44 x 1)]
antimony (Sb) = atomic weight 119 [14 + 17 + (44 x 2)]
bismuth (Bi) = atomic weight 207 [14 + 17 + (44 x 4)]

We note several things about Dumas's list. First and most obvious, unbeknownst to him, he had listed the elements in the exact vertical order of
Group V of the periodic table. Using the same reasoning but finding different mathematical patterns, Dumas uncovered a relationship between the elements F, Cl, Br and I (Group VII); O, S, Se and Te (Group VI); and Mg, Ca,
Sr and Ba (Group II).[26] (See fig. 5. 3.) He had stumbled onto the vertical order of three more groups.

But also note that if we use our contemporary, far more accurate atomic
weights and try to set up a Dumas-like mathematical pattern, things become
a bit fuzzy, to say the least.[27]

---

[26]Dumas mistakenly listed lead (Pb) after Ba because of his highly inaccurate atomic weight of
lead.
[27]Especially since, in using modern atomic weights, we are taking into account the different
isotopes of the same elements.

1 H																	2 He
3 Li	4 Be											5 B	6 C	7 N	8 O	9 F	10 Ne
11 Na	12 Mg											13 Al	14 Si	15 P	16 S	17 Cl	18 Ar
19 K	20 Ca	21 Sc	22 Ti	23 V	24 Cr	25 Mn	26 Fe	27 Co	28 Ni	29 Cu	30 Zn	31 Ga	32 Ge	33 As	34 Se	35 Br	36 Kr
37 Rb	38 Sr	39 Y	40 Zr	41 Nb	42 Mo	43 Tc	44 Ru	45 Rh	46 Pd	47 Ag	48 Cd	49 In	50 Sn	51 Sb	52 Te	53 I	54 Xe
55 Cs	56 Ba	57 * La	72 Hf	73 Ta	74 W	75 Re	76 Os	77 Ir	78 Pt	79 Au	80 Hg	81 Tl	82 Pb	83 Bi	84 Po	85 At	86 Rn
87 Fr	88 Ra	89** Ac	104 Rf	105 Db	106 Sg	107 Bh	108 Hs	109 Mt	110 Ds	111 Uuu	112 Uub						

*	58 Ce	59 Pr	60 Nd	61 Pm	62 Sm	63 Eu	64 Gd	65 Tb	66 Dy	67 Ho	68 Er	69 Tm	70 Yb	71 Lu
**	90 Th	91 Pa	92 U	93 Np	94 Pu	95 Am	96 Cm	97 Bk	98 Cf	99 Es	100 Fm	101 Mb	102 No	103 Lr

Figure 5.3.

nitrogen (N) = atomi weight 14.007
phosphorous (P) = atomic weight 30.974 [14.007 + 16.967]
arsenic (As) = atomic weight 74.922 [14.007 + 16.967 + (43.948 x 1)]
antimony (Sb) = atomic weight 121.75 [14.007 + 16.967 + (45.388 x 2)]
bismuth (Bi) = atomic weight 208.98 [14.007 + 16.967 + (44.5015 x 4)]

Strangely, the kind of mathematical pattern that Dumas sought actually does exist, but it does so in relation to a level of subatomic order that he could not have known because it had yet to be discovered: the level of the proton and electron. If we line up the same elements according to the number of electrons in each of their shells (with each subsequent shell being further from the atomic nucleus than the previous one, much as each subsequent planet is further than the previous one from the sun), a very exact and distinct mathematical pattern emerges.

nitrogen (N) = number of electrons 2 + 5
phosphorous (P) = number of electrons 2 + 8 + 5
arsenic (As) = number of electrons 2 + 8 + 18 + 5
antimony (Sb) = number of electrons 2 + 8 + 18 + 18 + 5
bismuth (Bi) = number of electrons 2 + 8 + 18 + 32 + 18 + 5

There are two points of interest here. First, obviously Dumas was actually right in his assumption that nature had some deep, underlying order that conformed to a simple mathematical pattern. It was just deeper than he suspected. Second, the level on which he could work at the time, the comparatively gross level of atomic weight, was an amazingly reliable stepping-stone to the underlying order—again, a sign of a kind of benevolent condescension in nature, not the kind of benevolence nor the kind of condescension one would find if we lived in an unintended universe.

Working with Dumas's insights, John Newlands (1837-1898) discovered something even more amazing by doing something incredibly obvious—obvious, at least, to a pattern-seeking creature. He simply lined the elements up, one after another, in order of increasing atomic weight, beginning with hydrogen. When he did, he noticed that every eighth element had similar properties. He called it the "law of octaves," named after the repeating patterns of notes on a musical scale. Just as with music—where if you begin with C and count up eight notes you land on a C again (an octave higher), or begin with G and land on a G again—so also with the elements. Thus, Newlands discovered that if he began with lithium (Li) and counted the next eight elements according to increasing atomic weight, he ended up at sodium (Na) which has chemical properties quite similar to lithium; or if he began at beryllium (Be) and counted forward eight elements, he ended on magnesium (Mg), which has quite similar properties to beryllium. If we look at our periodic table, blacking out the transition elements, we note that there are eight groups of elements and that, just as Newlands realized, the elements in each vertical row have notably similar chemical properties. There is, then, a kind of law of octaves, and this law defines the order of the periodic table. (See fig. 5.4.)

But Newlands also found out that sometimes when he did this, the properties of the known elements did not quite line up; indeed they seemed to be off by one, as if (to use the analogy of music) you began with the note C, counted eight notes and landed on D. In his original table of elements, he left such spaces blank, as if to say, there *should* be an element here but we haven't discovered it yet. If he had had the fortitude to keep them blank, he would have correctly predicted the existence of as-yet-undiscovered elements—and would even have guessed their weight and foretold their chemical properties!—all based on the pattern. Unfortunately, Newlands had a failure of nerve and decided to shift the elements over, filling up the blank spaces. The results soiled his marvelous insight because every chemist could see that the chemical properties did not then line up and form any kind of pattern.

And now for a humorous cosmic conjunction. When Newlands lined up the elements by increasing atomic weight, he did the very human thing of tidying things up by numbering them as well, for this allowed easy counting according to the law of octaves. Hydrogen (H) was 1, lithium (Li) 2, beryllium (Be) 3, boron (B) 4, carbon (C) 5, nitrogen (N) 6, oxygen (O) 7, fluorine (F) 8 and so on. Since we are pattern-loving creatures, we like to keep things orderly. Little did he know that he had stumbled on, without seeing it, atomic numbers: that is, the number of protons in a given element. Recall that as we scan the first few horizontal lines, or periods, of the periodic table, each element is arranged in accordance with increasing atomic weight, beginning with hydrogen. But more importantly, each element is numbered by its atomic number—the number of protons in the nucleus of each atom. In the series just mentioned, we can see that Newland's numbering is off by one. That is because helium (He), with atomic number 2, had not yet been discovered. If we add helium, then lithium is actually third, beryllium fourth and so on. If he had stuck to his guns, his ascending numbers would have

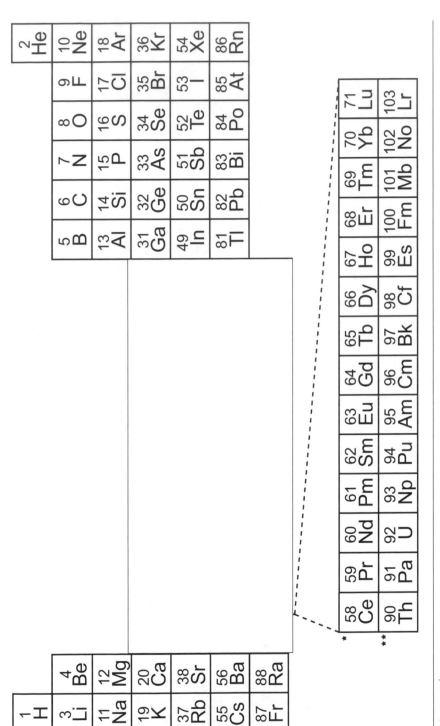

Figure 5.4.

prefigured the numbers of protons in each atom, before anyone had so much as dreamed of protons!

What a remarkably whimsical relationship between Newlands's intentions—merely to number his elements according to atomic weight so he could easily count off octaves—and the actual number of protons in each element, its atomic number on the periodic table! One can imagine being the designer of the table and feeling a bit like a human father who has cleverly hidden an Easter egg and is watching his children searching diligently, uncovering first one clue and then another, getting closer and closer and then temporarily grinding to a halt, flummoxed—all the while the surprise is hidden right under their noses.

It was left to the great Russian, Dmitri Mendeleev, to actually seize the egg, so to speak, to take the next bold step, for which he is credited with having finally cracked the code of the periodic table of elements (at least in regard to its general structure). Whereas Newlands had used the octaves of music as his guide, Mendeleev shuffled through cards. On each card, he put the chemical element and all its known properties, including the way it combined with hydrogen and oxygen. As is the case for pattern-loving creatures, he could not believe that nature would contain simply a pile of unordered elements, like so many randomly strewn playing cards. Surely there must be some elegant relationship. He worked and reworked the cards and noticed that, when they were arranged according to increasing atomic weight, a pattern began to emerge—the chemical properties recurred periodically, that is, regularly. That led him to place them in vertical columns, which he labeled Group I, Group II and so on, up to Group VIII. Believing that nature loves patterns as well, Mendeleev refused to allow that the pattern of increasing weights and chemical properties would be disrupted. Thus, when there was no known element that had the requisite properties to fit into a particular group or when there was a suspicious leap in atomic weight, he boldly left a blank. Even more boldly, he predicted that elements would be discovered to fill these blanks and, also, that they would have so-and-so properties and such-and-such atomic weights. If we compare his table to our current periodic table, it is wonderfully accurate (especially if we correct for the confusion caused by the transition elements).

And Mendeleev did indeed prove to be a prophet of discovery. For example, he predicted the existence of an element between calcium (Ca) and titanium (Ti), giving it the name *eka boron*, described its properties and assigned it an atomic weight of 44. A few years later, in 1879, Lars Nilson, a

Swede, discovered it, but he dubbed the element scandium (Sc) after his native Scandinavia. It has an atomic weight of 44.956. In the same way, Mendeleev successfully predicted the discovery of gallium (Ga) and germanium (Ge). These weren't just wild guesses. Mendeleev's deep chemical knowledge of the elements allowed him to predict a multitude of properties with stunning accuracy. For example, in regard to germanium (Ge), which he prenamed *eka silicon,* he quite accurately predicted the atomic weight, specific gravity, atomic volume, valence, specific heat and boiling point.

Mendeleev also corrected his contemporaries. For example, the accepted atomic weight for indium (In) at that time (76.6) would put it between arsenic (As) and selenium (Se), but Mendeleev saw that placing it according to the accepted weight broke the normal pattern of increase. He also noted that its actual properties did not fit the pattern of chemical properties that held elsewhere on the table. According to the chemical properties, reasoned Mendeleev, indium should go in between cadmium (Cd) and tin (Sn) and therefore must have an atomic weight of about 115. That is exactly where we find indium today (at. wt. 114.82). He likewise corrected uranium (U) and, interestingly, gold (Au). Mendeleev did make some errors, but he didn't fall into the trap of preferring his tidy little system to reality. He believed that more and more refined experimental work would continue to bring into focus the real order he argued was there. His intuitions, and his predictions, proved correct. Thus Mendeleev is properly credited with finally revealing to us the order of the periodic table of elements.

There is much more adventure in the history of the periodic table's discovery. Mendeleev lived before the revelation of the subatomic structure of the elements, that is, before the discovery of the deeper causes of the structure of the periodic table became known. How astounded he would have been if he could have witnessed with J. J. Thomson (1856-1940), Ernest Rutherford (1871-1937) and Niels Bohr (1885-1962) the discovery of electrons, protons, neutrons and the subatomic structure; or with Bohr, Erwin Schrödinger (1887-1961) and Werner Heisenberg (1901-1976) the quantum mechanical account that explains the order and characteristics of the elements even more beautifully and elegantly.

Of course, such a leap is impossible—and not merely because human beings cannot travel forward in time. The quantum mechanical account gives us a far deeper and more comprehensive understanding of the nature and structure of the periodic table, but it is both far removed from the world of the senses and from the much simpler mathematics used by Mendeleev. It offers greater intelligibility, but it makes proportionately greater intellectual demands—demands that only become known in the arduous and incremental work of scientific discovery. The remarkable thing is that nature condescends to set so many well-placed steppingstones in our path, from the world of sense to the quantum mechanical world. Without these, humans would never have discovered the order of the periodic table.

As rewarding as further analysis of the history of the discovery of the table would be, we must bring this chapter to an end. This brief overview allows us to underscore several key points. First, the very activity of scientific discovery is unintelligible if we take a reductionist view of human beings. The persistent drive to know the chemical order began not with the practical drive for self-preservation, but with the quite impractical desire for beauty. This desire for beauty was transformed, in alchemy, into the desire to create the most beautiful of elements, gold. And, as we have seen, perhaps the greatest intellectual leap of all occurred with Boyle's insistence that the chemical order is worth knowing for its own sake, no matter the costs or benefits. Whatever the practical benefits of chemistry, the desire to uncover the order of the elements, just for the sake of knowing it, was anything but practical.

Second, essential to scientific progress was a kind of continual correlation between the unshakable conviction that there is some underlying order in nature waiting to be discovered and the actual presence of an underlying order in nature. As we have seen, nature is not only ordered, but ordered in a kind of tutorial fashion, so that we, the knowers, can move from what is knowable in our everyday, visible, tangible experience, downward, step by careful step, through layers of previously unseen order, to the deep order we grasp only intellectually. That movement, that march, describes the march of chemistry through history, and the tutorial ordering in nature that made it possible couldn't be the result of an indifferent and pointless cosmos. One level of accidental order could be the result of chance; multiple layers of integrated order, configured in a way that is strikingly amenable to discovery, implies conspiracy.

If we find out *through* scientific discovery that the universe is intricately ordered in a way that invites discovery, then it's most reasonable to cease trying to imagine ourselves as the hapless creatures of a nihilist cosmos. As the history of chemistry reveals, when we reflect on ourselves as knowers, it is clear that we are pattern-seeking and pattern-finding creatures, creatures curiously made to be curious amidst an order curiously designed to be sought.

As we have seen, the chemical order proves itself to be remarkably math-friendly, and this, again, in an impressively tutorial way. We are relatively large creatures, and an enormous magnitude of size difference separates us from the atomic and subatomic levels of order. How strange that there should be any connection between the elegant and imperceptible subatomic world on the one side and, on the other, the comparatively clumsy, earthy, human-size activity of measuring the bulk weights of elements and sorting them according to their sensible qualities. How remarkable that the simple mathematical patterns that Dumas proposed, even though they were ulti-mately inaccurate, were leading rather than misleading. How extraordinary that something so ordinary as Newland's numbering the elements by in-creasing atomic weight should mischievously duplicate the atomic number, the number of protons. How remarkable that, in the end, the mathematics of the table is quite simple, simple enough for a beginning chemistry stu-dent, simple enough to explain in outline in a book such as this. Yet this simplicity is a simplicity of effect not of cause. The periodic table as a mas-terpiece is masterfully arranged and elegant, but that elegance is the effect of enormously complex and difficult causes (as we shall see more clearly in the next chapter).

In reflecting on the history of chemistry, this distinction between the causes of the chemical order and its effects leads to a fifth point. Human beings live on the level of effects and strive to discover the causes. For most of the history of discovery, humans were quite ignorant of the causes. They did not know that there were protons, electrons, neutrons and so on; they did not even know which elements were truly elemental; they did not fathom the connections between elements. But the effects, if attended to very carefully, led to the causes. Premature theoretical guesses about the causes of chemical order are both necessary and very human. But those guesses had to be tested against the actual order; and where they were held to dogmatically, they became a stumbling block rather than a help. We hu-man beings do not create the order, so our grandest philosophical claims, no matter how attractive we find them, must be humbly submitted to the order of nature. The opposite of such humility is the pride of dogmatism, the defiant cry of *non serviam* to the order of nature. When the order of na-ture contradicts the materialist's philosophic assumptions, it is just as mis-guided for him to hold to dogmatic materialist reductionism as it is to hold to the four-element theory of Aristotle or the theory of phlogiston. Such a materialist approach may have proved fruitful in the history of chemistry, but so too did the four-element theory, the theory of phlogiston and, for that matter, alchemy. When we tend carefully to the evidence of nature, it has a way of telling us when it's time to leave an old paradigm behind. Whether we choose to listen is another matter.

Finally, since the history of chemistry makes clear that humans-as-knowers rightly reason from effects to causes, then there is nothing intrinsically wrong with arguing from the ingenious order of nature to the Genius of nature. If we are not the creators of nature's intelligibility, if our intellect is indeed subordinate to that order, then we are obviously not the cause of nature's intelligibility any more than we are the cause of nature itself. That intelligibility is built into nature and, as we have noted at many points, built into nature in a tutorial way that seems to be ingeniously accommodated to human beings as students of nature. The chapter that follows will show how that Genius extends to the farthest reaches of the universe and to the beginning of time, converging on a pale blue dot in a vast cosmic sea, a curious little planet called Earth.

6

A COSMIC HOME
DESIGNED FOR DISCOVERY

There are two ways of getting home;

and one of them is to stay there.

The other is to walk round the whole world

till we come back to the same place.

G. K. Chesterton, *The Everlasting Man*

IN THE LAST CHAPTER WE FOCUSED ON HUMANS AS KNOWERS—in particular, on the history of the science of chemistry as evidence against materialist reductionism. The origin and activity of science contradict the reductionist description of human beings as merely bodily creatures defined by bodily needs and drives. Also contra materialism, the order of nature itself answers to the peculiarly human desire to know the truth about things for its own sake. A materialist would say we are guilty of anthropomorphism, of projecting our all-too-human needs on a blank and indifferent canvas, of seeing purpose in a pointless universe. Actually we are guilty of anthropism, not anthropomorphism, a guilt shared with an increasing number of scientists, for the universe and our place in the universe are guilty of anthropism as well. Indeed, during the last several decades, evidence has been accumulating that the universe is finely tuned not only for our existence, but also to allow us to discover the genius of the universe, including the discovery that it is seemingly designed for discovery.

Mathematically discoverable regularities lie at the depths of nature and at its origins, and they are configured in a way that allows us both to exist and to discover them. In cosmology, the term *anthropism* is used primarily in regard to the initial conditions of the universe that allow for complex carbon-based intelligent life to exist. Since human beings are the only complex carbon-based intelligent life of which we are aware (certainly the only technological life), these conditions have been called anthropic (from the Greek *anthrōpos,* human being).

According to its weakest form—called the weak anthropic principle—if certain things hadn't been as they are, then we wouldn't be here to notice them; but we are here, so those certain things must have been as they are. In its original form, as set forth by Brandon Carter, it has the flavor of a riddle: "What we can expect to observe [as scientists, in particular astronomers and physicists] must be restricted by the conditions necessary for our presence as observers."[1] In a later, somewhat more developed form, as stated by cosmologists John Barrow and Frank Tipler, it declares,

> The observed values of all physical and cosmological quantities are not equally probable but they take on values restricted by the requirement that there exist sites where carbon-based life can evolve and by the requirement that the Universe be old enough for it to have already done so.[2]

The discovery of so many anthropic conditions in the last half of the twentieth century (primarily in the last quarter-century) brought discussions of anthropic conditions to the center of cosmological debates and led to various other formulations of the anthropic principle.[3]

But given these two formulations, the question that should arise in the reader's mind is, why isn't the anthropic principle a kind of comical truism, like "Wherever you find yourself, there you are," or "If it weren't for me, I wouldn't be here"? The answer is rather complex, but goes to the heart of this book: it was a surprise in the late twentieth century to find that the universe was anthropic only because during the previous four centuries the universe came to be seen as *dis*anthropic.[4] And as we noted earlier, such

---

[1] Brandon Carter, "Large Number Coincidences and the Anthropic Principle in Cosmology," introduction to *Confrontation of Cosmological Theories with Observational Data,* ed. M. S. Longair (Dordrecht: Reidel, 1974), pp. 291-98.

[2] John Barrow and Frank Tipler, *The Anthropic Cosmological Principle* (Oxford: Oxford University Press, 1986), p. 16.

[3] We are laying aside, for the moment, all the other forms the anthropic principle has taken. The reader may refer to Barrow and Tipler's introduction (ibid.) for a fuller account.

[4] It has become conventional to refer to the "Copernican principle," that due to Copernicus's "discovery" that Earth was not the center of the cosmos, the cosmos could no longer be considered human centered, or anthropic. We resist using this term because, first, it glides over the actual complexities of the history of science, both in regard to Copernican and Ptolemaic astronomy; and, second, the intellectual source of the displacement was philosophical, originating in the cosmology of Epicurus who lived nearly eighteen centuries before Copernicus.

disanthropism ended in the misanthropism of philosophical and cultural nihilism, nihilism rooted in the modern revival of the ancient materialist cosmology developed by Epicurus. According to this view, the universe itself
was eternal, and everything in it was brought about by the random motion,
collision and adhesion of matter—matter defined as unbreakable and eternal atoms, differing only in size, shape and position. This universe was disanthropic because humans were in no way special, merely one of countless
accidental atomic accretions resulting from the random motion of indifferent, purposeless matter as it shuffled relentlessly through infinite space in
infinite time. Since humans were just one of an infinite number of accidents,
there was no purposeful connection between the movement of atoms on the
one hand and the presence of human beings on the other.

Over the infinite ages, so Epicurus argued, the universe spit out an unimaginable variety of creatures, both intelligent and unintelligent, on a multitude of worlds under an infinite number of conditions. At the core of disanthropism were, then, a set of assumptions: first, that the cosmological
fundamentals—the universe itself and its ultimate constituents, atoms and
the void—were eternally existing brute givens (and since they could not
have been otherwise, they needed no explanation, especially no explanation in terms of a creating intelligence); and, second, that homogeneous atoms themselves could exist in countless combinations, giving rise to innumerable forms of intelligent life under an endless number of conditions. As
scientific discoveries have now confirmed, both of these assumptions at the
heart of disanthropism were false.

The historical height of disanthropism was not in ancient Greece, but
rather in the modern West from the eighteenth century through the first half
of the twentieth century.[5] For scientists of the eighteenth and nineteenth
centuries, it was almost conventional wisdom that intelligent life was as
common throughout the universe as salt in the sea. It was often taken for
granted that every star was a sun that hosted planets teeming with intelligent
aliens and, even more, that these suns themselves were populated by the
most exalted and intellectually advanced aliens. This was not only a straight
mathematical deduction from the number of stars crowding telescopes in

---

[5]Readers should note that disanthropism doesn't necessarily entail atheism: many of the most
wildly enthusiastic believers in extraterrestrial life were Christians who believed that God, being infinitely powerful and infinitely beneficent, filled every available nook and cranny of the
universe with life. For a short account of this, see Benjamin Wiker, "Alien Ideas: Christianity
and the Search for Extraterrestrial Life," Crisis, November 2002, pp. 25-30.

ever greater numbers as optical power increased during this period; it was also taken as a reasonable inference from our own solar system.[6] In the nineteenth century the most eminent scientists and philosophers (and their long train of followers) waxed both confidently and eloquently about the existence of Solarians on the Sun and Lunarians on the Moon and Venusians, Jovians, Saturnians, Mercurians, Martians, Uranians and even cometarians elsewhere.[7] If our humble star so buzzed with complex, intelligent life on all its planets, so must virtually every other star in the cosmos. The belief in intelligent life on Mars was conventional wisdom among leading scientists as late as the first part of the twentieth century.

But nature did not long remain silent in regard to such deductive disanthropic optimism, and soon revealed a series of anthropic surprises. It was taken to be an obvious fact in the nineteenth century that the universe was eternal—so obvious that it came as a shock in the first half of the twentieth to discover that the universe had a beginning in the big bang and, furthermore, that there had to be very particular and peculiar fundamental fine-tuning at the universe's origin so that the big bang could be a big bloom.

Nor did the difficulties for disanthropism end there. Beginning with Lawrence Henderson's classic *The Fitness of the Environment: An Inquiry into the Biological Significance of the Properties of Matter* (1913), scientists began to realize that matter was not indifferently related to biological form; rather, the elements were discovered to be peculiarly fit for biology as it occurs on Earth. Since then, our greatly increased knowledge of the peculiar fitness of the elements for our biology has decisively undermined the disanthropic belief that homogeneous atoms could exist in countless combinations that would give rise to innumerable forms of intelligent life under innumerable conditions. Atoms are not homogeneous but quite distinct and carefully ordered, allowing only certain kinds of combinations under certain kinds of conditions. Further, certain elements (e.g., carbon) and certain compounds (e.g., water) were found to be optimal—not just a little more fit than other elements or compounds to function biologically, but incomparably more fit,

---

[6]The foolishness of a strict deduction from the number of stars to the number of alleged intelligent aliens manifests clearly the limitations of mathematics. Such heady calculations, however mathematically correct they might have been in themselves, must be tested against reality. It was our ignorance of the actual complex of conditions necessary for complex life that led eminent scientists to assert that there must be intelligent aliens in places where, as we now know, no life could exist.

[7]For the strange history of the belief in intelligent life, see Michael J. Crowe, *The Extraterrestrial Life Debate, 1750-1900* (Mineola, N.Y.: Dover, 1999).

completely eclipsing any other candidates. Against disanthropism, these discoveries have convinced most astrobiologists that our carbon and water biology is probably the only one possible in our universe for complex life.

As science advanced in the twentieth century, it also uncovered more and more examples of prebiological conditions that had to be met before any life—let alone intelligent life capable of a scientific revolution—could exist. If we add to all of these the various conditions necessary for a science of chemistry, then the possible chemistries consistent with our existence narrows even more dramatically. The science of chemistry has revealed itself to be deeply anthropic: it is seemingly formulated both for the existence of humans and to be discoverable by humans on a planet very much like Earth. Scientists, philosophers and theologians alike are grappling with what to make of this growing body of evidence and, once again, thinking about the universe in terms of what the Greek philosophers referred to as *telos*—that is, purpose.[8]

In his groundbreaking work *Darwin's Black Box*, Michael Behe argued that there exists "irreducible complexity" on the microscopic level, molecular "machines" that cease to function when even a single piece of machinery is removed.[9] What we shall find when we consider anthropism in the widest context is that a kind of irreducible complexity applies cosmologically, existing not only at the microscopic level of biology, but at the origin of the universe, in the characteristics of our planet and in between. We find, as it were, act on act of what we might also call organic complexity, unfolding

---

[8]The question being asked isn't "Is the fine tuning for technological life the sole purpose of the cosmos?" The question is rather, "Is it *a* purpose of the cosmos?" Thus, this thinking is not anthropocentric. Nor were the scientists and scholars before modern materialism anthropocentrists in any strict sense. If thinkers in the mainline of Christian orthodoxy, for instance, were to locate the central purpose of the universe anywhere, it would have been on God, arguing as they did that the universe is vast to properly reflect the majesty of its Maker. As noted above, there's actually a good deal of error that has accrued to the misnamed Copernican principle. Before Copernicus, the Earth was viewed not as the center of the universe so much as it was viewed as the bottom of the universe, with the center of the Earth depicted as hell, the farthest point from heaven. Contemporary depictions of pre-Enlightenment thinking on this matter are nearly as misleading as the flat-Earth myth, which depicts the educated of the Middle Ages as believing in a flat Earth, which they most certainly did not. See Dennis Danielson, *The Book of the Cosmos* (Cambridge, Mass.: Helix, 2000), and Guillermo Gonzalez and Jay Richards, *The Privileged Planet: How Our Place in the Cosmos Is Designed for Discovery* (Washington, D.C.: Regnery, 2004), chap. 11. Contemporary design theorists, in distinguishing their scientific work from any theological convictions they may hold concerning the purpose of the cosmos, argue that scientific evidence suggests that advanced life is an apparent purpose of the cosmos, and they emphasize that they make no pretense of having scientifically detected all the purposes of the cosmos.

[9]Michael Behe, *Darwin's Black Box* (New York: Free Press, 1996).

like an ingeniously organized play in which all of the physical constants and conditions are crucial to the unfolding story, reaching from the origin of the universe to the very peculiar conditions on Earth that have allowed for our existence and, further, for the scientific revolution itself. This complexity is irreducible insofar as all the conditions have to be met before science could occur. There are, in fact, so many conditions that an exhaustive discussion of them would fill hundreds of pages. But even a few will make clear why this intricate cosmic drama can properly be called irreducibly complex and why the so-called weak anthropic principle is no longer weak.

To begin at the beginning, in contrast to the belief of disanthropism, we have found that the universe did not exist eternally but began in an unimaginably hot and dense "point," expanding from this initial condition to what we see today. (Of course, in regard to the history of science, we first observed the expansion in the early part of the twentieth century and then inferred the origin as corroborating evidence flowed in.)

But why, we might ask, was the bang a bloom? That is, why didn't it expand so feebly that it simply collapsed back on itself in a tired heap? Or, having sufficient momentum to avoid collapse, why didn't it simply blow itself to cosmic smithereens, as explosions do on Earth? Or having occurred at a manageable rate, why didn't it lead to a universe of largely undifferentiated hydrogen atoms?

The answer is that the initial conditions were extraordinarily fine-tuned; they were, by all appearances, ingeniously orchestrated. Too much initial energy (or too little material), and the big bang would have produced cosmic confetti, the expansion so overwhelming the force of gravity as to prevent the formation of stars. Since stars are the crucible of the elements, without stars there would be no periodic table, no planets, no life. Too little initial impetus (or too much material) and the bang would have ended in a crunch, as the cosmic material collapsed back into a massive heap, like so many mutually attracting magnets. Again, the result is no stars, no galaxies, no solar systems and no life. That fine line is very fine indeed. Cosmologists have calculated that the actual density of matter in the universe as measured against the critical density of matter (the exact amount necessary for the perfect balance in expansion between confetti and crunch) cannot have differed by more than one part in 10^{15} and still lead to a complex, life-friendly universe.[10]

---

[10]Martin Rees, *Just Six Numbers: The Deep Forces that Shape the Universe* (New York: Basic, 2000), p. 99. The ratio of actual to critical density is routinely referred to as Ω by cosmologists.

We find the same kind of extraordinary fine-tuning in the constitution of
the elements themselves. Again, for both ancient and much of modern atomism, atoms were viewed as simple, homogeneous blobs, differing only
in size, shape and position. But the science of chemistry has made clear that
atoms are not homogeneous but are quite distinct, differing because of the
various and increasingly elaborate subatomic structures (as we move up in
atomic number). All of the elements are made up of protons (the positively
charged particle in the nucleus), electrons (the negatively charged particles
surrounding the nucleus) and neutrons (particles with mass but no charge
in the nucleus of all atoms but hydrogen). The elements are designated according to the number of protons each contains, with the various elements
also differing widely in their number of neutrons and electrons.[11]

In regard to the protons, it should strike us as odd that more than one
positively charged particle could reside in an atom's nucleus. After all, we
know from our own experience with magnets that like charges repel. Since
each element after hydrogen (running horizontally on the table) has one
more proton in the nucleus (helium has two protons, lithium three, beryllium four and so on), how is it that we have any atoms larger than hydrogen?
The answer is, the strong nuclear force, the force that overcomes the electromagnetic repulsion between protons. If this force were decreased by
about 50 percent, then we would end up with a very small periodic table
since the nucleus of atoms much bigger than hydrogen would simply fly
apart (or more precisely, never come together to begin with).[12] A decrease
of a bit more and your periodic table would consist of hydrogen alone.
While the possibility of a one-element periodic table might be some conso-
lation to the struggling chemistry student, it would have the unhappy effect
of eliminating the student himself.

Again, this is but one of many essential fine-tuning conditions. For life to
occur, all of the so-called fundamental forces in nature—the strong nuclear
force, the weak force, gravity and electromagnetism—must be finely tuned
not only in themselves, but in relationship to each other. For example, if the
electromagnetic force had been slightly stronger, then (as with a decrease in

---

[11]What distinguishes one element from another is the number of protons. The number of electrons will be the same as the number of protons if the atoms have a neutral charge (but ionized
atoms in the interiors of stars which have lost their electrons still retain their elemental identi-
ties). An element can also have multiple isotopes, which differ in the number of neutrons.

[12]On these figures, see esp. the excellent discussion by Robin Collins, "Evidence for Fine-
Tuning," in God and Design: The Teleological Argument and Modern Science (London: Routledge, 2003), pp. 178-99.

the strong nuclear force) our universe would have significantly fewer elements beyond hydrogen than it actually does.

Consider the even more precise fine-tuning necessary to beget two among the heavier elements essential to life, carbon and oxygen. The strong and electromagnetic forces must be finely tuned to each other. A change of more than $1/2$ percent in the strength of the strong force or of 4 percent in the electromagnetic force would destroy the balance in the number of carbon and oxygen atoms in the universe. These forces, in part, determine the finely tuned nuclear resonance of carbon that allows stars to "build" carbon from helium and beryllium, and oxygen from carbon and helium.[13] Since carbon and oxygen are the building blocks of the larger elements, the rest of the periodic table would not arise without them.

And just as with hydrogen, it is not only a case of our having a much shorter periodic table. Biological macromolecular complexity is impossible without carbon and oxygen; hence, without these chemicals, no chemists. Physicist and agnostic Fred Hoyle, who predicted the necessity of this nuclear resonance, remarked in regard to such extraordinary fine-tuning, "A commonsense interpretation of the facts suggests that a super intellect has monkeyed with physics, as well as chemistry and biology, and that there are no blind forces worth speaking about in nature."[14] How odd that we would have to go so far to recover common sense.

We note that this "commonsense interpretation" precludes the kind of monkeying around made famous in the typing monkeys canard and, indeed, in Darwinism itself. The fine-tuning occurs prior to any possibility of significant random chemical activity, for it precedes the production of the chemical elements, and hence it is prior to the "laws" of chemistry arising from their particular subatomic structures. This fine-tuning obviously cannot be the result of natural selection. Natural selection, the "mechanism" that Darwin hoped could displace the need for intelligent design, can only work among living organisms having the capability of reproducing and handing on beneficial traits to their progeny. But there is no such survival of the fittest possible in the prebiological realm of the earliest stages of the uni-

---

[13]To be more exact, the cascade involves, first, the production of beryllium from two helium atoms (^{1}He + ^{1}He → ^{8}Be) and then carbon from beryllium and helium (^{1}He + ^{8}Be → ^{12}C). Oxygen is then forged from carbon and helium (^{12}C + ^{1}He → ^{16}O).

[14]Quoted in Paul Davies, *The Accidental Universe* (Cambridge: Cambridge University Press, 1982), p. 118. Fred Hoyle's famous quotation comes from an unpublished University of Cardiff preprint entitled "The Universe: Some Past and Present Reflections."

verse—there is only the existence of the fittest, that is, the existence of un-imaginably precise fine-tuning.

For those wishing to sidestep the implications of the carbon resonance, things go from bad to worse as we examine the additional elements neces-sary for life. While the fine-tuning in regard to the production of hydrogen, oxygen and carbon is amazing, more is needed for biologically complex be-ings capable of science. The biological complexity of a human requires twenty-seven of the elements on the periodic table. This makes the job of fine-tuning the laws and constants of nature all the more dicey.

But even if, somehow, we could exist in a universe with just the twenty-seven elements in our body, such a poverty of elements (a bit less than one-third of the naturally occurring elements on the periodic table of elements) would make the science of chemistry, and much of the physics it led to, im-possible. As should be clear from our discussion of the discovery of the or-der of the periodic table, Mendeleev and those scientists immediately pre-ceding him needed over twice the number of elements that occur in our body for them to discern the pattern of the table. The richness of elements on Earth as a laboratory, a richness that goes beyond the requirements for life, allowed us to fill in the periodic table.

For example, there is beryllium (Be), which is essential for the stellar syn-thesis of carbon, or uranium (U), which fuels the inner furnace of our planet through radioactive heat allowing Earth to be a living planet.[15] But even if we add together all the elements that are necessary both for the human body and for a world capable of supporting a human body, we still have an ex-traordinary richness, a superfluity, of elements on Earth. Without these "su-perfluous" elements on Earth, it's unlikely we would have been able to dis-cern the proper order of the elements as found neatly displayed on our periodic tables. Happily, Earth is a surprisingly well-stocked laboratory.

We cannot take this rich supply of elements for granted. The distribution

---

[15]This heat drives volcanism and plate tectonics. While volcanic explosions and tectonic plate shifts are deleterious up close, they are essential to continent building. Without continents, we would live in a water world, where the multiplicity of elements would not be circulated through continent building and erosion but end up as a morass of silt on the Earth-sized ocean floor, and where (to say the least) it would be difficult for any intelligent life—if such could even exist—to discover and use fire so that they might uncover the chemical elements and their order. Volcanism and plate tectonics are also essential in the work of recycling car-bonates. They are subducted by plate tectonics back into the Earth's crust and there pressure-cooked. This releases carbon dioxide, which eventually makes its way back to the surface through springs and volcanoes. Gonzalez and Richards describe this carbon cycle and how it helps regulate the Earth's climate in *Privileged Planet*, pp. 55-57.

of elements in the universe is lopsided, in regard to both time and place. In regard to time, as we've noted, the elements build sequentially, from smallest to largest. According to the most recent research, the first stars may have formed as early as 155 million years after the big bang, but these stars (called Population III stars) were metal free, that is, free of any elements heavier than helium. (Astronomers refer all the elements beyond the first two, hydrogen and helium, as metals.) Obviously, such stars provide a short and sweet periodic table, but not a very fruitful one, and they certainly could not support life.

While not fruitful in and of themselves, these stars became the furnaces that created the next generation of stars, the so-called Population II stars. These are the oldest visible stars—from 10 to 13 billion years old—but they are quite metal poor; that is, they are made up of 99.99 percent hydrogen and helium. These metal-poor stars, in turn, gave rise to Population I stars, metal rich stars like our Sun. In regard to cosmic time, then, Population I stars are the youngest of the stars, all of them forming within the last 10 billion years. Only Population I stars (which consist of about 2 to 3 percent metals) are rich enough in metals to make planets and, hence, rich enough in elements to provide the complex chemistry necessary to build human beings and to supply a sufficient laboratory for chemists.

But even among Population I stars, metal content varies, with the oldest having about one-tenth the metal content of our Sun.[16] Stars with the metal content of our Sun are generally young to middle-aged Population I stars. (The most recent Population I stars can have $2\frac{1}{2}$ times the metal content of our Sun.) In regard to the expanse of cosmic time from the big bang (approximately 14 billion years ago), the provision of sufficient elements for a planet to have a well-stocked laboratory of elements, sufficient for the discovery of the periodic table, has only been possible (to make a generous calculation) within the last 5 to 8 billion years—only about half of cosmic time.

But much the same is true for cosmic place. We might think, staring up into the vast expanse of the heavens, that the science of chemistry is possible anywhere in the universe (just as we might have been tempted to think that it could have occurred at any time). But the galactic halo and the outer portion of the galactic disk contain stars too poor in metals to form planets

---

[16]The first stars to form with enough metals for building planets like our own were located in the dangerous inner regions of their home galaxies, so they were probably uninhabitable. See ibid., pp. 181-82.

rich in a variety of elements. If you were inclined to believe in intelligent
extraterrestrial life and were interested in communication, you would have
to aim your efforts at a relatively small part of the cosmos, at spiral galaxies
and, more specifically, at a broken ring in the disk of galaxies called the galactic habitable zone, about halfway between the radiation-bathed center
and the metal-starved edge and between the dangerous arms of the spirals.
Not only would indigenous complex life be essentially impossible outside
such galactic habitable zones, but (assuming the impossible for a moment)
if there were intelligent life elsewhere, such "aliens," bereft of a sufficient
laboratory of chemical elements, could never develop chemistry and, hence,
could never develop the technology to send and receive messages across
interstellar space.

Yet the conditions that make laboratory Earth so rich are even more strin-
gent. Just because you have a middle-aged or younger Population I star, like
our Sun, does not mean that you will find a number of well-stocked planetary laboratories circling it. In regard to our own solar system, the number
of elements present at or very near the surface on Earth is extraordinary, not
ordinary, as compared to the situation on other planets. Earth has an amazing 3,000-plus mineral species in its crust. Only about one hundred are
known to exist on other planets in our solar system, with Mars (at a distant
second) having the next most.[17]

The elemental richness of Earth as a laboratory is due in part to its being
the solar system's only living planet; that is, it is a planet with a combination
of volcanism, plate tectonic activity and an active hydrological cycle—all of
which contribute to bringing the suite of elements to the surface, concen-
trating the great variety of elements in mineral ore deposits where they can
support life and where they are available for investigation by curious creatures. Recall the importance of gold in the history of chemistry. Gold would
not be concentrated near the surface were it not for our fortunate combination of geologic and hydrologic activity, and the same is true for all the other
beautiful ores that enticed the earliest alchemists to dig into nature's chemical treasure trove. As geologist George Brimhall notes, an exacting set of
peculiar conditions was required:

> The creation of ores and their placement close to the Earth's surface are the
> result of much more than simple geologic chance. Only an exact series of

---

[17]See ibid., chaps. 3 and 5. The figures concerning the number of minerals were provided by
Gonzalez in a correspondence with the authors.

physical and chemical events, occurring in the right environment and sequence and followed by certain climatic conditions, can give rise to a high concentration of these compounds so crucial to the development of civilization and technology.[18]

Without these conditions having been met, Earth would not only be a much poorer laboratory, but of course, since life demands a ready supply of a variety of minerals near the surface, such paucity would mean that Earth was uninhabitable. If we compare Earth to the other planets even within our own quite habitable solar system, we find that the other planets, lacking some or most of these conditions, are not only bereft of life, but would have an insufficient number of available elements in mineral form to allow for the science of chemistry—in short, no chemists and no chemistry.

Indeed, the case against disanthropism is stronger when we factor in more and more of the conditions necessary for not only complex life, but for the kind of beings who can observe, discover, and launch a scientific revolution. As noted, astrobiologists now realize that the possibility for organic complex life is restricted to a small percentage of galaxies and, further, to a thin band within such galaxies known as the galactic habitable zone (GHZ) and, even further, within a particular star system's circumstellar habitable zone (CHZ), a narrow band also referred to as the Goldilocks zone—not too hot and not too cold but just right.

The ancient Greek philosopher Epicurus and many of the most prominent modern scientists in the eighteenth and nineteenth centuries, ignorant of so many of these conditions, assumed without any direct evidence that the possibility of intelligent alien life was so great that it was not merely probable, but certain. As science has advanced beyond mere speculation to actual evidence since the mid-twentieth century, our knowledge of the ever greater number of conditions that must be met before complex, intelligent life could exist is shrinking the window of probability to an ever-smaller circle.[19]

For instance, we circle a star that is neither too big nor too small, neither too hot nor too cool. Additionally, if we were allowed to go shopping around the galaxy for the star with the best type of electromagnetic radiation for organic life, our Sun would be hard to beat. Our Sun is the golden mean, a

---

[18]George Brimhall, "The Genesis of Ores," *Scientific American*, May 1991, pp. 84-91.

[19]For a more thorough analysis, see Michael Denton, *Nature's Destiny* (New York: Free Press, 1998); Peter Ward and Donald Brownlee, *Rare Earth* (New York: Copernicus, 2000); and esp. Gonzalez and Richards, *Privileged Planet*. Our debt to Gonzalez and Richards, esp. in this section, is great.

metal-rich G star, providing us with the necessary elemental materials, but it's not too metal-rich. If a star is too metal-poor, it cannot even make planets; but if the original "birth cloud" out of which a star formed is too metal-rich, then it produces a planetary system that is a disorderly jumble of planets and comets colliding, tugging and flinging planets out of the habitable zone.

In contrast to such chaos, our system has nine planets—not too many but very nearly. They are on surprisingly good terms: an improbably high number of planets with nice, well-behaved and very nearly circular orbits. We are often misled by the history of science to think of the orbits of our solar system's planets as strikingly elliptical because one of the great advances in astronomic accuracy, by Johannes Kepler (1571-1630), was the discovery that the orbits were ellipses and not perfect circles. What is lost in this account is a more accurate, larger cosmic perspective in which we note how amazingly close to circular these orbits really are. Astronomers measure orbits by how much they deviate from a perfect circle, a measure they call the orbital eccentricity (using a scale running from 0 to 1, where a perfect circle is 0 and a runaway ellipse is 1).[20] The orbits of Venus and Neptune are the most circlelike, with eccentricities of 0.007 and 0.010, respectively. The orbital eccentricity of Earth is 0.017, a less than 2 percent deviation from circularity.

Interestingly, the wildest orbits are those of the closest and farthest planets, Mercury (0.206) and Pluto (0.248). If you were to "stand above" our solar system, you would see the orbits of Mercury and Pluto as elliptical, but the orbit of Earth would appear circular. That near circularity not only ensures that planets aren't smacking into each other or disturbing each other's orbits, but that we on Earth enjoy a relatively narrow range of temperature fluctuations. A significantly more elliptical orbit would very quickly drive us above or below the narrow biological temperature range.

The smaller the temperature range needed, the more difficult it is to attain and maintain. It is not just that Earth is consistently the right distance from the Sun, which keeps it so wonderfully within the thin biological temperature range. Chief among the conditions that control Earth's temperature is our atmosphere. Without an atmosphere, Mercury has no "cover" protecting it from the full radiation of the Sun and no life-giving oxygen to allow for

---

[20]A circle has one focus, or center. An ellipse has two foci. Given the same circumference as a circle, the farther apart the foci are, the flatter and more elongated the ellipse becomes, approaching ever closer to simply being a straight line in which the two foci lie, as endpoints, on the "longways" circumference of the ellipse (or more technically, the major axis).

complex metabolic activity. Venus has the wrong kind of atmosphere, made up of over 95 percent carbon dioxide, a greenhouse gas that traps the Sun's radiation. The result is that, even though Venus is twice as far from the Sun as Mercury, it is a scorching $860°F$—two hundred degrees hotter than Mercury. None of the other atmospheres in our solar system are even remotely close to being able to support advanced organic life.[21]

Our extremely life-friendly atmosphere hinges on a complex of unlikely conditions. It is in every sense a biosphere. Nitrogen is one of the "big four" elements: nitrogen, hydrogen, oxygen and carbon are the core elements essential to biological complexity. Proteins are essential for complex life, and proteins are built from amino acids, the very name coming from the amino group (H_2N) that constitutes part of all amino acids. The four nitrogenous bases (adenine, guanine, cytosine and thymine) are the information bearers of DNA. The nitrogen is made available to higher organisms through bacteria that "fix" nitrogen organically, making it biologically available to plants, which, in turn, make it available to animals.

The extraordinary amount of oxygen in our atmosphere both protects us and provides energy for the activities of complex life. Ozone (O_3) protects us from the Sun's harmful ultraviolet radiation (O_3 being produced when O_2 molecules in the atmosphere are broken up by ultraviolet radiation). Then, of course, animals need oxygen to breathe. Astrobiologist Guillermo Gonzalez and philosopher Jay Richards offer one of several reasons why oxygen was the right element for the job:

> Life relies on chemical energy for its immediate metabolic needs, and chemical energy is all about the exchange of electrons. The most energy is released when elements located on opposite ends of the periodic table exchange electrons. Oxygen is second only to fluorine in the amount of chemical energy released when it combines with other elements.[22]

Even more remarkable, the amount of oxygen in the atmosphere not only allows for complex life, but is also the result of complex life. The most obvious and immediate explanation for our anomalously high level of oxygen on Earth is photosynthesis, the process in which plants, taking in the energy of the Sun, convert atmospheric carbon dioxide and water into their carbon-based sugar structure, with the byproducts being water and oxygen. Plants, therefore, not only mediate nitrogen to animal life, but also provide its at-

---

[21]Mars comes closest, apparently possessing the ability to host at least subsurface microbial life.
[22]Gonzalez and Richards, *Privileged Planet*. p. 37.

mospheric oxygen, which both protects living things from the Sun's radiation and makes possible animal respiration. Again, our atmosphere is particularly difficult to obtain and sustain without the existence of an active
biosphere. It's a chicken-and-egg problem: how do you get the proper atmosphere without the biosphere, and how do you get the biosphere without the proper atmosphere? Atmospheric scientists offer some highly speculative proposals, but these proposals further constrain the early conditions
necessary for life to have gained a foothold and persisted over long periods
of time.

Equally fortunate, the greenhouse gases so crucial to keeping Earth's temperature within the biologic range constitute a self-regulating thermostatic
system. When there is too much CO_2 in the atmosphere, the Earth heats up.
When it heats up, plants grow more luxuriantly (both from being in a hot-
house and from increased CO_2) and greater evaporation occurs. But plants
breathe CO_2, thereby removing it from the atmosphere, and the greater
evaporation leads to greater precipitation, which also removes the excess atmospheric CO_2. Things cool down. If things get too cool, precipitation decreases, and less CO_2 is removed, allowing atmospheric buildup to occur
again from volcanic activity and animal respiration. Without this finely tuned
thermostat continually running, Earth would be uninhabitable.

There are other facets of our atmosphere that make not only life, but also
science possible. We take for granted that we can *see* things, and sight is our
most important sense for science. If our planet were covered with a very
dense atmosphere, things here on Earth would be quite dark, a world of
grays and blacks at best. At the same time, we need enough of an atmosphere not only to provide energy for living things, but to protect us from
harmful light. The electromagnetic spectrum is vast, reaching from low-
energy, long-wavelength radio waves to high-energy, short-wavelength
gamma rays. Without an atmosphere, we would be prey to the whole spectrum, and the higher energy gamma, x-ray and ultraviolet wavelengths are
so energetic that they destroy the chemical bonds that make biological complexity possible. Life on a planet needs protection, and such protection must
come, for the most part, from the atmosphere. Yet too much protection
would mean blocking out visible light, that small band of very useful electromagnetic radiation nestled between ultraviolet and infrared. Blocking out
visible light removes one of the obvious conditions for seeing.

That makes having the right atmosphere a rather daunting design problem. As Michael Denton points out, the range of visible light on the entire

electromagnetic spectrum "represents the unimaginably small fraction of approximately one part in 10^{25} of the entire electromagnetic spectrum—equivalent to one playing card in a stack of cards stretching halfway across the cosmos, or one second in 100 quadrillion (100,000,000,000,000,000) years."[23]

Now the sign of a designing genius is the ability to provide a solution to a seemingly intractable problem—a solution that displays the best overall fitness, a "part" that not only solves the problem, but also functions simultaneously as the solution to other design problems within the same system. This is what we find with our atmosphere. The ozone (O_3) and oxygen (O_2) in our atmosphere—which exist largely because of Earth's plant life and which make possible the existence of higher, more complex animal life—also serve to block out the most lethal electromagnetic gamma rays, x-rays and ultraviolet rays; and because of their peculiar chemical properties, quite suddenly open up a "window" allowing in the very thin spectrum of visible and near visible light. Similarly, the carbon dioxide and water vapor, present in the atmosphere in their respective amounts because of the quite peculiar characteristics of Earth, block out some, but happily not all, infrared light. Moreover, photosynthesis uses an even smaller slice of this very thin slice of visible light to create the oxygen. A substantially more energetic wavelength would make photosynthesis or any biological activity impossible, and any less energetic wavelength would be too weak to bring about chemical reactions sufficient to fuel biology.

While perception beyond visible light is possible in the near ultraviolet and infrared regions of the spectrum that exist on either side of visible light, the maximum window of our atmosphere occurs at visible light. And contra Darwinism, our planet's fauna could not have evolved eyes differently to take advantage of whatever type of electromagnetic radiation was available. Visible light is optimal for visual perception according to the parameters of the chemical bonds that make complex biological life—with complex organs of sensation—possible. In regard to the eye, visible light is energetic enough to bring about the needed chemical reactions in the eye and of small enough wavelength to allow for sufficient visual discrimination. As we trail into the more ultraviolet light, the higher frequency, while allowing greater discrimination, causes more and more biological damage; as we trail into the infrared, the lower frequency makes for poorer discrimination and less energy,

---

[23]Denton, *Nature's Destiny*, p. 51.

which, even though less dangerous, makes chemical reactions too weak.[24]

Perhaps we might imagine that our atmosphere could have been life-friendly but covered in clouds, blocking our view of the distant stars.[25] Fortunately it doesn't. We must add to all this that our Sun is strangely fit to provide the proper wavelength. Unlike most other stars, our Sun emits 40 percent of its electromagnetic radiation in the visible range, peaking exactly in the middle of the visible spectrum.[26] In sum, we don't just have a pretty good situation for vision; we have one optimized for both life and scientific discovery.

But our atmosphere serves science in another way. As we have seen in the last chapter, the importance of fire cannot be overestimated in the history of chemistry. Without it, pure metals (with the exception of gold) could not have been extracted from ore, nor could the alchemists have heated liquids and distilled from them their constituent elements. Fire, to us, seems as natural as the air we breathe, but that is because we take for granted the conditions that allow for such fire, including the air we breathe. We have fire—tamable, usable fire—because of a complex of conditions.

Fire needs oxygen—the right amount of oxygen. Too little, and you cannot have fire; too much, and you have either runaway fire or an explosion. The Earth's atmosphere now has the right amount, just under 21 percent.[27] It has been estimated that for every 1 percent increase over this amount, the probability of ignition of a forest fire increases by 70 percent.[28] But if oxygen were to be decreased, large mammals like us would begin to struggle. The level of atmospheric oxygen, then, sets a narrow condition for life and science; and the Earth's level of atmospheric oxygen, the result of a complex of conditions, ingeniously satisfies several design criteria simultaneously.

Fire beneficial to civilization and science requires not only this optimal level of oxygen, but a second kind of fuel as well. The fire that results must be controlled, not explosive—suitable to warm oneself, cook food and en-

---

[24]To grasp the full power of these connections, see ibid., chap. 3.

[25]In fact, this is an unlikely possibility, given that dense cloud cover would upset other biologically necessary conditions.

[26]Gonzalez and Richards, *Privileged Planet*, p. 66.

[27]Scientists now think that the oxygen content of our atmosphere peaked at around 35 percent about 200 to 250 million years ago, just before the Cretaceous period. Such high levels mean that forest fires from lightening strikes would have been frequent and especially devastating, if not for some mitigating factors (e.g., the presence of generally higher humidity). Human beings as firemakers were not on the scene at this point. By the time of our arrival, oxygen levels were at or near present-day levels.

[28]See J. E. Lovelock, *Gaia* (Oxford: Oxford University Press, 1987), p. 71.

gage in chemistry. For this, you have to have organic matter to burn such as wood, coal or oil. The ability to control combustion is essential not only for keeping warm and cooking our food, but also for chemical activity and analysis, from its earliest forms in smelting to the Bunsen burner of the modern laboratory. Such combustion is controllable because of the relative inertness of carbon, the key ingredient in wood, coal, natural gas and oil fuels.

This is no small point. We are all familiar with the science-fiction landscape, showing a desert planet with its indigenous master species, exotic tripeds shuffling across the barren surface from their alien futuristic housing to their shiny metallic flying saucers. The imaginative portrayal is meant to manifest the disanthropic assertion that some radically different type of planet could easily host advanced, technological life. But this is nonscientific fantasy. The existence of such nice, shiny precision metal parts for flying saucers depends not only on the availability of ores near the surface, but on the slow advance of technical expertise in the smelting of metals. This in turn depends on having the means to perform the kind of controlled burning that metal smelting requires—the kind of controlled burning that organic matter, especially trees, provide. But the alien landscape is utterly barren of vegetation. This barrenness, we must understand, entails both a technological barrenness and an intellectual barrenness. Advanced technology of the kind portrayed in such science fiction is only possible with the advanced understanding of the order of the chemical elements. No matter what planet you are on, no matter where you are in the universe, the chemical elements themselves and their order do not appear on the surface, but, as should be clear from the last chapter, are uncovered only after a long, slow and winding analysis using an incalculable amount of fuel. In short, if a planet is not vegetation-rich like the Earth, then ascribing advanced technological life to it is groundless fantasy. (That is doubly true, by the way, when we remind ourselves that having significant oxygen in our atmosphere is the result of having photosynthesizing plants.)[29]

Realizing all this helps to dispel the silliness of pseudo-science fiction that all too often masquerades as truth grounded on firm scientific principles.

---

[29]Continuing to bracket off for the moment obvious habitability problems, one could posit a bubbly planet where the fire derived from volcanic activity, but this would restrict the scenario to geologically active planets, and the lava would create a chicken-and-egg problem: it could provide a source for fire, but one would require a certain technological savvy to put hot lava to practical use, a skill that would need to wait until the inhabitants had spent centuries using fire in a highly controlled manner.

Humans have often had visions of radically different, indigenous, intelligent life far more advanced than ours, existing on planets radically different from our own. Such visions were (and still are) put forth, even and especially by scientists and science enthusiasts, and their enthusiasm makes them "see" what isn't really there. At the dawn of the early twentieth century, for example, Percival Lowell, peering through a telescope, trumpeted the discovery of advanced canal work on Mars, a certain sign, so he argued, of technological life.[30]

The difficulty with such science fiction—even beyond the fact that the wonderfully precise, geometrical canals "seen" by Lowell and others didn't exist—is that the very thin atmosphere of Mars has far too little oxygen (about 0.13 percent) to support either complex life or even the humblest of fires necessary for technological life.[31] It also lacks the vegetation that makes carbon available for controlled combustion. But without both fuel and fire, there can be little technological progress and certainly no progress in the science of chemistry. Once again, the conditions existing on Earth demonstrate how narrowly constrained are the possible conditions for science.

So much more could be added to all of the above: Earth's particular axial tilt, spin rate, and mass; the beneficial stabilizing effects of our unusually large moon; the Earth's radiation-shielding magnetic field; the tectonic activity that makes volcanic eruptions possible and builds continents; and a growing list of other prebiotic conditions that make Earth fit for life.[32] Again, with each added condition that we find necessary, the more difficult a living planet is to attain; and the more difficult it is to attain, the more difficult it is to maintain over significant periods of time. If we add to all this the extraordinary fine-tuning that allowed for the synthesis of oxygen and carbon to begin with, we see more and more that the anthropic law is focused ever more narrowly on creatures like us.

But there is yet another consideration that narrows this focus even more. Knowing means going beyond the surface and examining the full complexity and potentialities of the thing under consideration. A casual, surface

---

[30]Lowell did not find the so-called canals. That honor goes to Italian astronomer Giovanni Schiaparelli, observing Mars with a modest telescope in 1877. But Lowell was the most adamant and effective popularizer of this scientific fiction. See esp. Steven J. Dick, *Life on Other Worlds: The Twentieth-Century Extraterrestrial Life Debate* (Cambridge: Cambridge University Press, 1998), pp. 26-43.

[31]See Gonzalez and Richards, *Privileged Planet*, pp. 85-86.

[32]For a fuller account see Ward and Brownlee, *Rare Earth*, and Gonzalez and Richards, *Privileged Planet*.

reading of William Shakespeare's works is not sufficient to know the works' real depth. The same is true of the chemical elements. We know an element most thoroughly and exactly by understanding the actual expression of its full potentialities as they are manifested not only in isolation, but in the greatest chemical complexity. That is, we know carbon best when we see its powers expressed in all possible forms, from the simplest compounds to the most complex carbon-based life. If you want to know what the full potentialities of the element carbon are, you would want to be immersed in the maximum complexity of carbon-based life. Or to see it in the negative, if you want to know what hydrogen is capable of, don't go to a hydrogen planet like Jupiter where just about all you have is hydrogen. Stay on Earth and see it at work in DNA and a multitude of protein structures.

Similarly, you won't find the fullest expression of the powers of iron on the iron rust-red planet of Mars; you'll find it better in the hemoglobin of living blood coursing through our veins. We in our habitat are, taken together, the best specimens of the full drama of chemistry, especially when we note that, from Earth, not only we are able to find the full potentialities of chemical elements actualized in our immediate environment, but we are also able to read the chemical composition of the most distant stars—and this because the very transparency of the chemical constituents of our atmosphere provide an optimal window to let in visible light and also an optimal window for gazing out on the beauty of the heavenly host.[33] In sum, Earth is the paragon of laboratories for the paragon of animals. In this sense, both the chemistry of our world and the science of chemistry are anthropic.

It should be clear, then, that the weak anthropic principle (WAP) is weak indeed; that is, its weakness is its disanthropic assumptions. Materialists use it to argue that the conditions discussed above aren't surprising because we wouldn't be around to notice them if everything weren't fine-tuned for our existence. WAP, as it is most often used by the disanthropically inclined, is meant to enable a shrug of indifference to anything remarkable we see, a big "So what?" in the face of the most ingenious of natural features.

But this disanthropic shrug of indifference ill fits an anthropic cosmos. A sign of its essential weakness is that the stronger the evidence is *against* disanthropism, the more self-congratulatory becomes the shrug. As we have seen, there is not only fine-tuning of the initial conditions ("Yeah, so what?"); all the way forward, we find that our existence as technological creatures de-

---

[33]See Gonzalez and Richards, *Privileged Planet*, chap. 4.

pends on a building up of conditions in very precise ways in regard to our galaxy ("Yeah, so what?"), our solar system ("Yeah, so what?") and Earth's own constitution and atmosphere ("Yeah, so what?"). The disanthropic have placed themselves in an enviable position: no amount of evidence against it is evidence against it, for otherwise we wouldn't be here to discuss it.

Oxford philosopher Richard Swinburne likens such strange reasoning to the situation of a man before a firing squad who opens his eyes to find that the bullets missed him and formed a precise outline of his body on the wall behind him. Instead of immediately and sanely looking for some purpose behind the squad's precision firing that spared his life, "The prisoner laughs and comments that the event is not something requiring any explanation because if the marksmen had not missed, he would not be here to observe them having done so."[34] The squad would be right to wonder whether perhaps one bullet hadn't knocked the sense out of the prisoner. The peculiar event does demand an explanation, though not in terms of chance, but of purpose.

The problem with dodging an explanation through WAP is that it distracts attention from the anthropic evidence by focusing on the necessary conditions as if they were sufficient and, further, by treating the necessary conditions, without argument or demonstration, as a matter of cosmic luck. In that way, WAP skirts the demand for an explanation.

Even fine-tuning just at the cosmic level, considered apart from all of the fine-tuning discussed in this chapter, destroys the WAP's appeal to inevitability. A sign of the fatal wound is the attempt to circumvent the implications of cosmological fine-tuning by appealing to multiple, undetectable universes. According to multiverse theory, our universe is one of untold millions of universes, and the fine-tuning of our universe for life and discovery is the result of a precosmological selection effect: we happen to be in a fine-tuned universe that allows for our existence as complex, intelligent biological beings capable of science; otherwise we couldn't notice the fine-tuning.

According to this explanation, what we don't observe are the potentially infinite number of other universes, either existing before ours or existing in other dimensions, that blew up into lifeless bits, collapsed into lifeless rubble, produced only hydrogen, produced only heavy elements or even produced entirely different chemical elements. So while it seemed for a brief

---

[34]Richard Swinburne, "Argument from the Fine-Tuning of the Universe," in *Physical Cosmology and Philosophy,* ed. J. Leslie (New York: Macmillan), p. 171.

historical moment that prebiological fine-tuning trumped materialism, pre-cosmological materialism hopes to trump fine-tuning: "Of course we're in a universe consistent with our existence. You didn't expect to find yourself in one of the bazillion other universes that didn't allow for you, did you?"

It is difficult to deal with evasions such as multiverse theory; that is, the evasions are so evidently ludicrous it is difficult to take them seriously. A hypothetical alternate universe is, by definition, not causally connected to our universe and so not in any danger of being outed as a mere fiction. Unfortunately, since the desire to avoid the commonsense conclusion of design is so strong among some scientists, many have retreated to such Alice-in-Wonderland reasoning.

Modern science—as opposed to science fiction—has revealed to us only one universe, one that, contra Epicurus and his many followers, is not an eternal, brute given but one with a beginning that demands an explanation. It is a universe, moreover, whose exquisitely fine-tuned initial conditions—allowing as they do for the existence of scientific observers—have the hallmark not of givens but of ingenious gifts, a point that will become still more evident as we reflect on the genius of the elements in chapter seven.

7

THE GENIUS OF THE ELEMENTS

To see things in the seed, that is genius.

Lao-Tzu

THE CHEMICAL MAKEUP OF OUR WORLD IS INGENIOUS not merely
in some vague sense but in a very specific way: it is astonishingly fine-tuned
for life and science. More than this, it exemplifies the elements of genius
manifested by William Shakespeare and other great artists down through the
centuries. To see this, we take up the story some years after the puzzle of
the elements had at last been fitted together. While Russian chemist Dimitri
Mendeleev was able to establish, in outline and in significant detail, the periodic table of elements, it was another half-century before scientists uncovered the causes of the order on the subatomic level, the level of protons,
neutrons, and electrons.[1] Humans never would have discovered this subatomic level if adumbrations of its order did not reverberate to the surface
as clues that allowed the science of chemistry to proceed.

Recall how helpful the correlation between atomic weight and the order
of the elements was. Without this rough correlation, chemists could not have
laid out the elements in order; and without laying them out according to
atomic weight, no visible pattern of octaves, of elemental characteristics repeating themselves every eighth element, would have appeared. And these
octaves, in turn, hinted at so much more. Fortunately, while an imperfect
guide to the table's order, atomic weight was close enough to draw scientists, pattern-seeking creatures, into those greater depths of insight. Our repeated encounter with such good fortune more and more strongly suggests

---

[1]Scientists would discover, among other things, that the cause of the different atomic weights
was the number of protons, neutrons and electrons in a given element, and that the cause of
the triads' "lining up" was the number of electrons in the outermost energy level.

another pattern, one poorly explained on the grounds of materialism—namely, that nature is arranged for human explorers and that it is not so easily arranged that we would judge the prize of discovery of little worth, but is challenging enough that, having concentrated humanity's best efforts, we are able to move further up and further in.

The design of the periodic table, which is far from arbitrary, is based on the ingenious design and interrelationship of the elements themselves. The great chemists who discovered it are, then, not the primary subjects of this chapter. They are like cryptologists laboring to crack an ancient hieroglyphic and, having succeeded, find themselves reading a work of unsurpassed genius. Now, our culture throws around the word *ingenious* almost as thoughtlessly as it tosses around the once-majestic term *awesome.* Here the use of the term *ingenious* is very deliberate and specific. The periodic table of elements, in all its exquisite order, fits the qualities that we traditionally associate with the works of genius to the highest degree.

To begin at the end of the quest, the table possesses depth: it's dense, rich, subtle and difficult to exhaust—not only in itself, but as an alphabet for the drama of physical existence. Scientists have unraveled its mysteries all the way down below the level of the proton to the quantum level, and they are still digging. At the same time and for all its depth, the work is readable or, to use Thomas Aquinas's more suggestive term, it possesses clarity. The clarity, the brightness, is displayed here on two levels, one above the table and one within. Growing from the order of the elements, nature is not content to combine atoms in a dreary and colorless assemblage of larger and larger homogeneous blobs. Quite the contrary. By the time protons have coupled with neutrons, electrons have found their place in the dance of the subatomic cosmos, and the atoms have formed themselves into compounds—including the organic compounds of living things. The vividness of scene is enough to keep naturalists, poets and children at play in the fields for age upon age.

Reductionists have often used a faulty argument (the fallacy of composition) to convince us that the vivid reality of our everyday experience is unreal. They consider the color the poet celebrates a mere illusion, the red of the rose no more real than the rose itself. They argue that if we push lower to the true reality, atomic reality, we find it to be both entirely roseless and entirely colorless. In their rage for reduction, they snatch away every color along with the rose itself. But this is mere small-mindedness. Of course the atoms of individual elements aren't colored, since they are smaller than the wavelength of light, the medium by which color can be expressed; yet the elements taken together in compounds of sufficient quantity are colored, reflecting certain wavelengths and absorbing others. It's just as silly to say that a strawberry isn't red because none of the individual chemical elements is red as it is to say that a certain bridge won't support a car because none of the individual iron atoms could support a car.

Color isn't unreal; it is a quite real potentiality hidden in atoms waiting to unfurl, made possible by the elegant order of the chemical elements along with the proper conditions that allow color to be sensed, including (as we have seen) not only the many intricate aspects of vision, but also a planet with an atmosphere that lets in the spectrum of visible light while blocking out other lethal radiation. In the same way, the order of the chemical elements makes the beauty of the flower possible, even though none of the elements is a flower or can express the complex functions of a living plant. The red rose is quite real and quite red; and as a complex, living thing, it allows for the expressions of some of the deep potentialities hidden in the elements.

The elements possess clarity at another level as well, the clarity of their order as revealed in the periodic table. Once the young chemistry student understands why the table is ordered as it is ordered and sees shining forth the depth of intricately related insights that this leads to, she may feel as one gazing on Samuel Taylor Coleridge's pleasure dome of Kubla Khan, "a miracle of rare device," a realm bright with elegance on every level.[2] This is part of the growing evidence that the genius of nature is an anthropic work—a work that seems to have us in mind. Just as it was with the order of geometry, we see the way the order of the elements condescends to offer steppingstones to discovery.

On the grounds of philosophical materialism, this is a surprise. After all, according to physicist Steven Weinberg's understanding of the universe, the origin of chemical order, indeed of all order, is gibberish—chaos that somehow, through a chain of accidents, brought about the more than one hundred elements familiar to us from the periodic table. Even if we grant that some strange and impersonal ordering principle governs the chain of accidents, why should it be so solicitous and condescending as to make the result intelligible *to us?* The order of the elements and the elements themselves might have been hidden in nature so deeply, so far from our senses and our

---

[2]The phrasing here is from Samuel Taylor Coleridge's "Kubla Khan."

everyday existence, that no amount of digging, intellectual or practical, could have unearthed them.

To ask the same question from a different angle, why is the final order as manifested on the periodic table a simple enough text to read, so that a beginning chemistry student can grasp its essentials in a very short time? After all, even if we take for granted the discovery of the elements and their order, the final structure of the table could have been as complicated as learning quantum mechanics, so that a beginning chemistry student would face a seemingly impenetrable, meaningless maze of complexity.[3]

This is not a mere offhand comparison. The actual order of the table is dependent on the existence of discrete energy states on the quantum level. These discrete energy states allow for electrons to have definite energy levels in distinct orbitals surrounding the nucleus of the atom, and these in turn not only determine the vertical order of the table in groups, but also define how it is that atoms bond and react with one another.

Yet even though quantum mechanics is necessary to understand the ultimate causes of the table's order, the main lines of the table's order are accessible, fortunately, on a far simpler level. If it were necessary to master quantum mechanics to discern the order of the periodic table, it isn't that only quantum physicists could do chemistry: there would be no quantum physicists or chemists, because we could never have leaped the enormous gap from our everyday experience to the quantum level.

Now of course the order of the elements could have been simpler. "What's with this unwieldy Lanthanide series! And the Actinide scandal!" a frustrated chemistry student might exclaim. "A first-rate deity wouldn't let them dangle off the bottom of the chart!" But like the colorless paraphrase of Miranda's speech from *The Tempest* that appears in chapter three, such a narrowly considered tidying up of the table would have been at the sacrifice of nature's many shades and subtleties.

We said of Shakespeare's works that their clarity is not an abstract thing isolated from play and audience. The clarity directs the meanings of the plays to English-speaking humans. It was an anthropic clarity, one bent on impressing on the viewer as much of the play's richness as possible. What

---

[3]To get a sense of the importance of this point, taken from the vantage point of the human effort to capture the order, we may compare our present, wonderfully simple and concise table with the immediate precursors of Dmitri Mendeleev's version, such as the telluric helix of A. E. Béguyer de Chancourtois. See Aaron J. Ihde, *The Development of Modern Chemistry* (New York: Dover, 1984), p. 240.

English professor hasn't heard the whining complaint, "Why couldn't Shakespeare or Yeats or Browning have just *said* it?" Indeed, students are so sure that the poet should have made things much easier for them that they opt for the simpler version—the Cliff Notes. But what a falling off is this! The more fundamental question is, *should* it have been simpler?—a question worth asking about the elements as well.

The confusion can be traced back to a misunderstanding about what it means for a work of art (or a work of technological artifice) to be optimal. What we could call an optimal work of genius does not possess each quality of genius in some isolated maximal form, not for humans in this life. Actual works of genius possess what Henry Petroski calls constrained optimization.[4] Consider a family automobile. A good design will seek to maximize roominess, fuel economy, maneuverability, safety, quietness, acceleration, comfort, low price and a number of other features. One doesn't have to look at the list long to see that some of the criteria are in tension. Even in the absolute best family car, all of these qualities will not be maximized in isolation one from another.

Imagine if a rock 'n' roll star looked inside a doctor's family automobile, the roomiest and most luxurious family car made by Lexus, and shook his head. "What a lousy family car. My tour bus has beds, easy chairs, a wet bar. Your sorry car doesn't have any of this." We would rightly assume that this rocker was off his, for obviously the tour bus would make a lousy family car, and not just because of its extravagant cost and ruinous fuel economy. Imagine Mom trying to maneuver the thing around in the mall parking lot with her kids running up and down the length of the bus like little heathens.

Or consider a Shakespearean sonnet. In its series of three quatrains and concluding couplet, the rhyme scheme neatly mapping the flow of the poem's argument, it has more of what we might call mathematical elegance than does *Hamlet,* but the latter far outstrips the Bard's best sonnet in the qualities of depth, richness of harmony and even clarity rightly understood. Likewise, the order of the chemical elements is an ingenious negotiation of often-competing criteria. The order has enough surface clarity to allow humankind to unravel its mysteries so as to appreciate its deeper clarity, the

---

[4]Henry Petroski, *Invention by Design* (Cambridge, Mass.: Harvard University Press, 1996), p. 30. The introduction to Guillermo Gonzalez and Jay Richards, *The Privileged Planet: How Our Place in the Cosmos Is Designed for Discovery* (New York: Regnery, 2004), as well as several of the subsequent chapters, develop this concept regarding the way the Earth appears optimized for both habitability and for allowing us to make a range of scientific discoveries.

brightness of its inner workings. It's readable enough to be read, but it also demands that we stretch those capacities unique to humans.

Part and parcel of this, the order of the elements does not fling itself on the human intellect like a cheap work of pulp fiction. Understanding the general wisdom behind this, Shakespeare made the work of comprehension "uneasy . . . lest too light winning," the audience would judge "the prize light."[5] The chemists who unraveled the periodic table and lived to see it largely completed were in no danger of this. No one can doubt they possessed a heightened and deepened appreciation for the beautiful order they labored to uncover. And the contemporary chemistry student will value it all the more for its demanding something from her. In this sense of its possessing constrained optimization for our discovery, appreciation and intellectual edification, the order of the elements is deeply anthropic, exquisitely fitted both to stretch and satisfy human capacities.

One place where it may be difficult to see the constrained part of the optimization is in the periodic table's harmony and elegance, for here, there seems to be little constraint. The chemists and physicists did not find complexity without architecture, depth without design. The world of the atom opened up by the periodic table is one where nothing is merely tacked on; and many diverse phenomena can be described by surprisingly elegant descriptions, like the Bohr theory of the atom proposed in 1913, itself soon surpassed by another of greater elegance, though it still acts as a stepping-stone for beginning students.

Here we recall Coleridge's description of the imagination, particularly that of the imagination of genius—a creative force that "dissolves, diffuses, dissipates, in order to re-create; or where this process is rendered impossible, yet still at all events it struggles to idealize and to unify."[6] Coleridge felt that Shakespeare possessed such imagination to the highest degree. He might have recognized a spirit akin to Shakespeare if someone had revealed to him the relationships among the elements at their lower levels, relationships that permit the most various combinations and qualities, an order allowing the simplest atom to become part of water in one instance while in another to flame out in the burning brightness of a new star.

---

[5]Shakespeare *The Tempest* 1.2.448-50. Prospero is here speaking of the need to make Prince Ferdinand work to win Miranda's hand in marriage but, like so much in Shakespeare, the passage also works as a comment on artistic practice.
[6]Samuel Taylor Coleridge, "Biographia Literaria," in *The Oxford Authors: Samuel Taylor Coleridge* (New York: Oxford University Press, 1985), p. 313.

By combining these various qualities here noted, works of genius again and again afford the experience of surprise. The student of nature, like the student of Shakespeare, can expect surprises—delightful parallels, double meanings, illuminating motifs and connections of every sort, enough to keep a multitude of scientists happily occupied for generation upon generation. Thanks to the high artistry we find in the order of the elements, there is a seemingly inexhaustible trove of surprises, each hidden, yet not so far down it cannot be found.

And more than mere icing on the cake, the peculiar joy of surprise is fundamental to the experience of scientific discovery, permeating the scientist's entire being, body and soul. Two episodes from the history of chemistry illustrate this, the first occurring when the great chemist Sir Humphry Davy discovered the element potassium through electrolysis. As his brother John Davy reported, when Humphry

> saw the minute globules of potassium burst through the crust of potash, and take fire as they entered the atmosphere [pure potassium is highly reactive], he could not contain his joy—he actually bounded about the room in ecstatic delight; and some little time was required for him to compose himself sufficiently to continue the experiment.[7]

Similarly, when the rather staid giant of physics Sir Ernest Rutherford (1871-1937) directed a beam of alpha particles at thin gold foil and noted the pattern of deflections on the coating of phosphorescent material, it is reported that Rutherford, a native of New Zealand, burst into an impromptu celebratory haka, the war dance of the New Zealand Maori.[8] Why such exuberance? The deflections revealed a great surprise: the structure of the atom consisted in a very small positively charged nucleus surrounded by electrons at an enormous distance from the nucleus. If the gold atom's nucleus were the size of the dot on this letter *i*, the remainder of the atom—virtually all empty space—would be the size of a garage.[9] Even matter as hard and substantial as metal was practically all space, and Rutherford couldn't help but dance at the discovery. Such is the proper thrill of a rational animal, this strange creature of spirit and matter, intellect and body.

---

[7] Quoted in J. R. Partington, *A Short History of Chemistry*, 3rd ed. (New York: Dover, 1989), p. 184.

[8] See Cathy Cobb and Harold Goldwhite, *Creations of Fire: Chemistry's Lively History from Alchemy to the Atomic Age* (New York: Plenum, 1995), pp. 272-74.

[9] Ibid.

And this is a most revealing point, for like Shakespeare's works, the peculiarly human activity of discovering the elements reveals more than the elements. It reveals us to ourselves, and thus, it is anthropic in yet another sense. As with Shakespeare's dramas, it teaches us deep truths about human nature. The dramatic spectacle of chemistry seems to be staged for us, well apportioned as it is for human beings, creatures who are an essential union of rationality and animality. The science of chemistry is best suited to an intellectual creature who knows through the senses. This may sound like an odd claim to make because it seems obvious. But a little familiarity with the history of philosophy shows us that human beings are prone to two opposite errors, two confusions about their own humanity that cause them to lose their humanity: on the one hand, reducing themselves to mere animals, and on the other, elevating themselves to pure angels.

Chemistry teaches us otherwise. It is not a science that is possible for mere animals (like dogs).[10] This isn't merely because dogs aren't clever enough. It also goes back to our human appreciation for beauty, both physical and intellectual. Nor would we better off if we could "shuffle off this mortal coil" and study the world as purely intellectual, bodiless creatures. At the entrance to the theater of the elements, there is a sign: "No taste, no touch, no service." The underlying order of the table is not an abstract truth deduced from purely intellectual axioms, a priori principles or rationally self-evident laws. The table represents certain aspects of physical reality as made known through physical reality and, hence, through our senses. The ancient philosophers generally missed this point. Acting as if the underlying order was simply a set of speculative truths known by the intellect alone, they invariably fell into error, error that could only be corrected when they were willing, figuratively and literally, to get their hands dirty and to be brought "back to their senses" by the sensible physical order of nature.

The ancient philosophers' mistake persists in diluted form, even to the present, coming to us by way of the dualist philosopher René Descartes

---

[10]In saying this, we do not meant to imply that other animals are merely sensing machines, devoid of reason—an error of modern mechanistic materialism where an animal's actions are reduced to mechanical responses to sensitive stimuli, an error that cannot explain even the complexity of a dog's actions, let alone a human's. As even the ancient Greek philosopher Aristotle understood quite clearly, animals share, more or less, in a kind of capacity for judgment that relies on sensation but is not reducible to it. What animals lack, and human beings have, is the desire and capacity to make abstractions from the sensible, physical order to intelligible immaterial truths; hence animals can neither discover nor grasp the abstract relationships underlying the order of the periodic table. To them, the table can never have meaning.

(1596-1650), who depicted the human being as a sort of ghost in a machine, a spirit whose body is at best a tool and at worst a mere nuisance. Reacting against this view, many suffer today from believing that a human is instead a mere animal. It's salutary to see a very different view of the matter unfold from the history of chemistry itself, demonstrating the necessity of our entire being, bodies, senses and intellects in union. Chemistry is a sensuous science, a science that finds the truth through the senses rather than through purely intellectual construction and deduction. Its sensuousness is not an embarrassment—as later, colder scholars of Shakespeare thought of his ribaldry—but part of its rich anthropism; and without it, there would have been no science of chemistry.

All true chemists are therefore quite clear about the importance of our senses. The first thing that you would notice, for example, upon entering an alchemist's laboratory (indeed any active laboratory up until the invention of exhaust fans) was the smell. The diverse aromas are a key to the presence of certain chemical compounds and elements, as anyone who has ever smelled sulfur can attest, and chemical lists often include in their descriptions the smell of elements and compounds (especially insofar as they indicate the presence of a dangerous chemical). Detecting the connection between odors and chemical makeup has been essential in the history of chemistry, as chemists quite often had to sniff out clues to the order.[11]

The sense of taste? Like smell, it helped to reveal the underlying chemical order. One of the earliest systematic practical alchemical manuals, *The Secret of Secrets,* written by Abu-Bakr Muhammed ibn Zakariya al-Razi (850-c. 923), the famous and influential Persian physician and alchemist,

---

[11]Interestingly, Humphry Davy—the chemist mentioned in the previous chapter as so successful in "shocking" compounds apart through electrolysis to find hidden elements—got his start as a scientist inhaling gases as a research assistant in the newly founded Pneumatic Institution. The founder of the institution, a physician named Thomas Beddoes, believed that the many gases discovered by Joseph Priestley might prove beneficial to health. The unintended results of Davy's systematic sniffing—besides his near suffocation on several occasions—were the discovery of the giddying effects of nitrous oxide (laughing gas), the inhalation of which soon became a kind of national fad, especially among the young, and the consequent fame gained by Davy in publicizing its strange effects. Also note in regard to the sense of smell that there is even a name given to a particular kind of chemical radical of "fearful odor"—first discovered in 1760 and later isolated by the great chemist Robert Bunsen—called the cacodyl radical ($C_4H_{12}As_2$) from the Greek *kakōdēs*, "stinking," the name provided by another great chemist, Jön Jacob Berzelius (1779-1848). Also of interest, one of the four classes of hydrocarbons (hydrocarbons being the largest class of organic compounds) is the aromatics.

provided a useful and influential early classification of substances into metals, vitriols (any of a number of shining, crystalline substances), boraxes, salts and stones on the basis of solubilities and *tastes*.[12]

And one of the most important distinctions in chemistry, the distinction between an acid and a base, is also a matter of taste. Acids and bases (or alkalis, as the latter are also called) had been known for quite a long time, reaching far back into the annals of alchemy. The primary mode of classification, lasting up until about 1665 (when Robert Boyle discovered that syrup of violets turned red with acids and green with alkalis, the first "litmus test"), was that acids tasted sour (*acidus* in Latin means sharp or sour) and bases tasted bitter.

We've also seen that, oddly enough, it was the bitter taste on Alessandro Volta's tongue that awakened him to the possibility that alternating metals could be piled to make a continuous electric current. (The current would have been too weak to pick up by touch.) Without the voltaic pile, a host of elements would not have been discovered, nor would the essential electrical nature of the atom have been unveiled. Humbling? Not at all! It is revealing.

Also, in Boyle's early litmus test, color vision was essential. Boyle was also one of the first to recognize the importance of identifying the presence of particular chemical substances through the flame test. For example, he noted the green color given to a flame when copper salts were introduced.

This connection between chemical substance and color, begotten through fire, would prove of the greatest import not only for analyzing chemical compounds, but also for discovering them; it allowed us, through the science of spectroscopy, to decipher the chemical content of the most distant stars. Since each chemical element and compound has its own particular spectral signature in a device called a spectroscope, astronomers are able to discern the chemical ingredients of the most distant visible stars without ever leaving Earth. Without the sense of sight, and in particular our ability to distinguish colors, it would have been next to impossible to crack the code of the periodic table.[13] Without the spectral lines occurring

---

[12]Al-Razi is better known to the West as Rhazes.

[13]In this regard it is interesting to note that most mammals and colorblind humans are dichromats: that is, they have two kinds of functioning cones in their eyes, rather than three, thus restricting them from seeing the full range of colors in visible light. The great early nineteenth-century chemist John Dalton was colorblind, a limitation that would have handicapped him significantly if he had done chemistry in the second half of the nineteenth century, given the centrality of spectroscopy.

in the narrow slice of the electromagnetic spectrum where light is visible to the human eye, the elements discovered through spectroscopy may have remained beyond our reach, as would have the underlying pattern of the periodic table.

It bears repeating just how thin a slice it is. Again, as Michael Denton explains, it's a mere one part in 10^{25} of the full electromagnetic spectrum, "equivalent to one playing card in a stack of cards stretching halfway across the cosmos, or one second in 100 quadrillion (100,000,000,000,000,000) years."[14] Such an improbability is so vast as to boggle the mind.

A more tangible but no less amazing instance may help us sense the extraordinary convergence of factors. Our planet experiences not merely total eclipses, but perfect eclipses—where the Moon almost precisely covers the Sun, enough to block its dazzling direct light, allowing astronomers to observe its chromosphere. This oddly perfect "fit" of Moon over Sun is only possible because of the particular sizes, sphericity and distances of the Sun and Moon. The Sun is four hundred times larger than the Moon. But it's also four hundred times farther away, meaning it has the same apparent size on the face of the sky as does the Moon, and so the Moon just covers it during a perfect eclipse.

This is the first instance given in Guillermo Gonzalez and Jay Richards's book *The Privileged Planet* of the remarkable convergence of factors simultaneously conducive to life and discovery. The Sun and Moon are both good sizes and distances for a habitable planet. But what does this convergence of factors have to do with discovery? The extraordinary view of the solar atmosphere that occurs during perfect eclipses played a key role in unlocking the science of spectroscopy, the chemical makeup of the stars and, in turn, the order of the elements. And all this would have been lost to us without the sense of sight, and the intellect capable of discerning the connection between spectral colors and the presence of particular elements.

In sum, the periodic table wrung from us our utmost capacities even as it taught us about those capacities. This should give us pause. As with the scientists who discovered the periodic table, we ought to suspect—having found so many layers of anthropic order—that there is more to come. And indeed, more and more it looks like anthropism all the way down, in the sense that chemistry, physics and even cosmology all point to a universe

---

[14]Michael Denton, *Nature's Destiny* (New York: Free Press, 1998), p. 51.

fine-tuned for observers like ourselves, rational animals, creatures of sense and intellect.

This is nowhere more dramatic than in the way the order of the elements caters to our appreciation for beauty—both tangible and abstract. As we have seen, the sensible beauty on the surface drew human beings ever more deeply into the discovery of elements. And it was the belief in and search for an elegant, underlying relationship of the elements that fueled the labors of chemists, and this faith in nature's elegance paid off. Rather than finding homogenous matter or a mere heap of indifferently associated elements, scientists found highly organized atomic and subatomic complexity, a microcosmos that is not only elegant in itself, but again, ingeniously provides the material for all higher-level physical elegance. This culminates in the most complex level of all—the physical drama of everyday experience, the drama in which the full potentialities of the order of the elements are expressed: the world of color, sound, smell, taste and touch and of elegant form and elegant activity. In geneticist and botanist Edmund Sinnott's words, "Beauty's variety and profusion in sound and form and color . . . are far greater in the products of life than elsewhere. Beauty is of life's very essence. It is one of the permanent and indestructible parts of nature."[15]

And the appreciation of beauty, of the genius in the order of the elements, comes to fruition when we understand how exquisitely orchestrated that order is for the drama of life. Certainly one of the greatest surprises (and consequent thrills) in the last century is the discovery that, against the sober and humiliating tide of disanthropism, driven ultimately by the reductionist tenets of materialism, the universe is exquisitely anthropic, that is, exquisitely fine-tuned for life in the very finest details.

This comes into sharp focus when we consider the element carbon. Carbon is maximally fit for biology; to understand carbon is to understand it as the backbone of biological complexity. It's now generally assumed among astrobiologists that life has to be carbon-based because only carbon is fit to act as a backbone for the extraordinary chemical complexity that life necessitates. Carbon is not just a little more fit than its "competitors" on the periodic table, but maximally fit—an obvious sign of its chemically fruitfulness

---

[15]Edmund Sinnott, *Matter, Mind and Man* (New York: Atheneum, 1968), p. 132. Sinnott's analysis of beauty in nature, taking us far beyond what Charles Darwin could ever explain, is extraordinary (see pp. 127-35).

being that, of the 10 million chemical compounds known, 9 million of them are carbon compounds.[16]

One key to carbon's amazing maximal fitness is the ability to form carbon chains, that is, stable bonds between like atoms. While a few of the other elements on the periodic table can form such chains, only sulfur (S), tin (Sn), silicon (Si) and phosphorous (P) can form molecules of more than three like atoms. But S, Sn, Si and P are unfit for the demands of biochemical complexity because their chains are unstable and react with water and oxygen much more readily. The notion of non-carbon-based life, such as the often-suggested silicon-based life, is just one more example of disanthropic, non-scientific fiction. In compounds, carbon is stable but not too stable, and this is all to the good because life demands not only complex structures, but complex activities. If carbon compounds were too stable, they could not form and reform or act and react in the innumerable chemical processes that biological growth, metabolism and maintenance demand; if they were too reactive, everything would always be breaking down (or more likely, never be built up). But it is this wonderful balance that also constrains the conditions under which life can occur.

Keep in mind that astrobiology determined that carbon was necessary for life in spite of a prior prejudice against restricting the parameters for life elsewhere. So many things just fell into place for our existence. When we recall the amazing fine-tuning required at the origin of the universe to produce carbon, disanthropism becomes more and more irrational. Such disanthropic reductionism heads the wrong way. To understand carbon fully and essentially we have to understand its culmination in biology; the Greeks would say we must understand its *telos*. It is this maximal fitness for all of the chemical combinations and exchanges that occur in and among cells that teaches us to admire the elegance of carbon, its marvelous fitness for biology.

Carbon by itself, though, is in need of a suitable helpmate. On Earth it needn't look for long. A particularly delightful anthropic surprise, insofar as it was hidden under our noses all along, unsurpassed in its elegance, is hum-

---

[16]An entire branch of chemistry, organic chemistry, is devoted to the study of carbon compounds. It might seem that since it now covers all carbon compounds and not just those that occur in actual organisms, the name is a relic of the nineteenth century when chemical compounds produced from once-living matter were called organic (and all others inorganic). But carbon is so biologically fit that it is quite proper to understand carbon in terms of its biological goal.

ble and glorious water. It too is uncannily fine-tuned for life. From the singing voice of winding creeks to the roar of the ocean surf, poets have long celebrated its charms. But as inspirational as water has proved to the poet, and even to the early philosophers and later scientists who gave it the status of an element, only in the last century and a half has science revealed the extraordinary nature of water.[17] We have discovered that water is not just *a* liquid of life, but *the* liquid of life, expressing the same kind of amazing maximal fitness and ingenuity as carbon for organic life.

Consider just a few of water's unique properties. Like Euclid's superior form of the Pythagorean theorem, water is an ingenious stroke of elegant simplicity. A sure sign of designing genius is the ability to make something that is simple maximally fit for a multitude of tasks—and add beauty into the bargain. As we have noted, few chemical compounds are as simple as water, H_2O. But packed within that simple molecule are a host of properties, such that no other liquid rivals it for biotic fitness. This unusual set of properties has been called *anomalous* by recent scientists precisely because its suite of properties is without peer.[18]

We may begin with water's most famous anomalous property: it expands on freezing. As anyone who has taken the most basic of junior high science classes knows, when things get hot they expand, and when they cool, they contract. Simple physics. In regard to cooling and contracting, water follows this rule quite consistently, but only up to a point, that point being 4°C (39.2°F). Then suddenly, unlike almost any other substance, water *expands* as the temperature drops lower and in expanding, becomes less dense. The freezing process ends with a sudden burst of expansion as it freezes solid at 0°C (32°F).

This anomalous property proves essential to life. As the liquid of life, water must be readily available on the surface of the planet in large quantities.

---

[17]The first treatment of water's extraordinary properties was *Astronomy and General Physics Considered with Reference to Natural Theology*, written in 1832 by master of Trinity College, Cambridge, William Whewell. The next major treatment was Lawrence Henderson's minor classic *The Fitness of the Environment* (1913). Then came A. E. Needham's *The Uniqueness of Biological Materials* (1965), John Barrow and Frank Tipler's *The Anthropic Cosmological Principle* (1986), Harold Morowitz's *Cosmic Joy and Local Pain* (1987), Michael Denton's *Nature's Destiny* (1998), and most recently Philip Ball's *Life's Matrix: Water, a Biography* (1999). Each book in turn praises the amazing nature of water. There is now even a website dedicated solely to the ongoing collection and technical explication of water's extraordinary properties: <http://www.sbu.ac.uk/water>.

[18]The following analysis of water owes much to an unpublished paper that Benjamin Wiker wrote with Guillermo Gonzalez and Jay Richards.

If water *continued* to contract and become more dense right up to its freezing point, then as water froze on ponds, lakes and oceans, it would sink to the bottom and remain there untouched by the melting rays of the Sun. The floor of the body of water, critical to its biological cycle, would be interred in ice. And as each year passed, layer after layer of ice would build from the bottom up until little was left but great blocks of ice filled with frozen flora and fauna. Thanks to water's anomalous expansion, this death by ice does not happen.

This peculiarity of water is part of a larger set of unusual thermal properties, all of which contribute to water's outstanding fitness for life. Another is water's oddly high *specific heat*. If something has a high specific heat, it takes a lot of heat to change its temperature; if it has a low specific heat, then it will take little to heat it up. Water has the highest specific heat of any liquid except ammonia.

What difference does that make for life? If you've been to the beach, you probably felt the striking difference between the specific heat of sand and water: the sand nearly burns the bottoms of your feet, but the water cools them off deliciously. The same Sun hits both sand and water, yet the water absorbs the heat while remaining cool. But the importance of water's wonderful ability to absorb large amounts of heat goes far beyond keeping us cool at the beach. Water makes our planet habitable. About 70 percent of Earth is covered with water. To recognize the importance of this, imagine trying to live on a planet whose surface is nothing but scorching sand. Or if you don't live near the beach, simply stand in a parking lot on a hot day. To raise the temperature of liquid water one degree requires nearly five times as much heat as it takes to raise a similar amount of sand or concrete one degree.

The high specific heat of water also allows for temperature moderation inside animals. Anyone who has had the flu knows the debilitation that can be caused by a fever when body temperature rises even one or two degrees. If our bodies were not over two-thirds water, we would be burned up by the heat thrown off by our own metabolic activities. It has been calculated that if we subtracted the cooling effects of water within our bodies, the heat generated in one day just by our metabolic activity at rest would raise our temperature to between 100°C and 150°C.

But it isn't merely the high specific heat of water and the bulk water content of our bodies that enable us to live. It is another remarkable, related property of water, which the equally remarkable complexity of our body makes use of for its preservation—the extraordinarily high *latent heat of vaporization*. It takes a lot of heat to evaporate water; or to put it the other way around, when water evaporates, it takes an enormous amount of heat away.[19]

To take advantage of this property of water, we have a remarkable cooling system much more efficient than any engine's: perspiration. The greatest danger of overheating occurs when we are active and it is hot, and it is precisely then that we sweat most profusely, maximizing the water available for evaporation on the surface of our skin. Without water's high heat of evaporation and its uniform dispersion over our body surface, such careful temperature regulation would be impossible for us.[20]

This unusually high latent heat capacity both cools and warms the planet as well. The evaporation from the oceans absorbs an enormous amount of heat in the tropical regions. This latent heat is then carried to colder latitudes, where the water vapor condenses, changing back into liquid water. As it does so, it releases its cargo of latent heat—the same amount of heat originally absorbed in evaporation—and warms up the colder climates.[21]

Without this continual cycling of heat and cold, most of our planet would be either too hot or too cold for habitation. But because roughly three-quarters of Earth is covered with water, this liquid of life acts as a large temperature moderator, not only by absorbing heat, but transferring it, keeping Earth from either heating up too much or cooling down too much. This *prebiological* property of water provides a continuously self-regulating temperature system, making our planet habitable by keeping temperatures extraordinarily steady within the thin range of temperatures suitable for life. No other known substance could absorb, store, circulate and dispense this much heat.

---

[19]This heat is not lost but is whisked away and stored as latent heat. To raise a set amount of water from freezing to just boiling (0°C to 100°C, or 32°F to 212°F) it takes about 100 kcal of added heat. To turn that same water into water vapor (i.e., steam) will take another 540 kcal of heat—over five times as much heat as it took to bring the water to the boiling point. That's how much heat evaporation removes.

[20]We add "for us" because other animals have equally ingenious ways of dispensing with excess heat.

[21]This climatic mixing and moderating of temperatures occurs not just above the oceans, but (making use of water's high specific heat) in the water currents of the oceans as well. Surface currents are mainly driven by wind, whereas deep-water currents are driven by water-temperature differences. The effect of both, however, is that the great amount of heat absorbed by water in warmer climates is carried to cooler climates and released, and that the cooler water is carried back again to warmer climates.

That is not all, however. Water also has an unusually high *latent heat of fusion;* that is, just as water absorbs vast amounts of heat as it changes from liquid to gas, so also it absorbs great amounts as it changes from ice to liquid. How much? Well, its latent heat of fusion is more than that of iron.[22] This property also makes water a perfect fit for the liquid of life, especially in regard to global temperature moderation. When liquid water freezes, it *releases* the same amount of heat that it takes to turn the ice into liquid water. Thus, when you complain of the cold in mid-January as you watch a lake slowly freezing over, remember that it would be even colder were it not for the release of latent heat that occurs as water freezes.

We may now move from water's anomalous thermal properties to its power as a *solvent.* The power of water as the universal solvent makes it uniquely fitted as elixir of life (and for the various laboratory tasks of the chemist). More substances dissolve in water than in any other solvent, yet water (unlike, say, most acids) is not a highly reactive solvent. If it were highly reactive and corrosive, it would attack biological entities even while it was doing the helpful task of freeing minerals from rocks for use in living things.

We may now see the striking congruence of water's peculiarly fit properties, in particular the great importance of water's expansion near the freezing point and its power as the universal solvent. Without the continual flow of water over rocks, minerals necessary for life would stay largely locked up in nonbiological form. But water has a trick to unlock more of these precious minerals. It creeps into the cracks and crevices of rocks while still in the liquid phase and then expands on freezing, breaking the obstinate rocks apart to yield greater, more ragged surface areas that water can more readily dissolve when spring temperatures return it to its liquid form.

Water is also both a solvent and a circulator of life, on the planet and in the body. It breaks down and circulates crucial minerals throughout the globe and, likewise, breaks down nutrients from larger, more complex substances and transports them from one part of an organism to another via the blood stream. (Our blood is over 91 percent water.) Thus, the great global streams keep a host of precious compounds circulating around Earth—

---

[22]Of course, iron melts at a much higher temperature. The melting point of water is 0°C, while that of iron is 1808°C. But water's heat of fusion—the heat required to change 1.0 kg of a substance from the solid to the liquid state when its temperature is at the transition line—is slightly higher, with water requiring 79.7 kcal/kg and iron 69.1 kcal/kg.

potassium, sodium, calcium, magnesium, iron, silicon and many more—and the small streams within us continuously circulate life-giving proteins, sugars, fats and a variety of precious, life-sustaining minerals.

On both levels, this marvelous cycle depends on two *other* anomalous aspects of water. For water to act as the great circulator, it must be a *liquid in the biological range* and it must be of the right *viscosity.* A liquid that boils away before it ever hits the biological range, such as ammonia (the boiling point of which is −33.4°C, −28.12°F), would be just as unsuitable as a liquid that, while it didn't boil away, was too viscous for the arteries of living things (liquids such as olive oil, glycerol or pitch). We add that water is one of the very few substances, if not the only substance, that exists in all three phases—solid, liquid and gas—within the biological temperature range, the range within which carbon-based life can occur.

Could there be anything else striking about this most striking of compounds? Well, yes, as can be seen clearly in a drop of water. Water becomes a drop because of its anomalously *high surface tension,* higher by far than any other ordinary liquid and surpassed only by liquid mercury. It is this property that makes water surpass all other liquids in regard to its ability to pull itself up to greater heights.[23] The surface tension allows water to rise in the thin tubes, or capillaries, of plants, because the tension seeks a kind of equilibrium: it draws water up the capillaries as it evaporates out of the pores, or stomata, of plants, as if water were a long chain or a series of buckets on a rope. The high surface tension of water keeps the plant from having to exert energy pulling water up to its leaves.

Recalling what we said above, the stomata of leaves are cleverly designed to take advantage of water's anomalously high surface tension by allowing openings at the top of the plant through which water can evaporate. Were it not for water's unusually high surface tension *and* the presence of stomata, there would be no large vegetation and hence no chemistry. Why? Without trees, there can be no wood; without wood, controlled fire is difficult; and without controlled fire, there could not be chemistry. Interestingly, we can thank the unusually high surface tension of water for another form

---

[23]Additionally, water does not just stick to itself and, hence, pull on itself. It adheres to other things as well. This ability to adhere to many other substances is called "wetting." Water "wets" glass, cotton fabric, common rocks and nearly everything organic and inorganic in soil. In a narrow glass, such as a test tube, you can see the water adhere to the side, seeming to climb up the glass, making the surface of the water look concave. In plant capillaries, the pulling up of water via evaporation is aided by the adherence of water to the sides of the thin tubes themselves.

of beauty as well—rainbows. This glorious prism effect is only possible because water is able to form large, round drops.[24]

Interestingly, if we look below Earth, we find another property of water beneficial to growing things. Why doesn't rain just soak through the soil and flow straight down into deep underground streams before the thirsty roots can drink any of it? The answer is that water adheres to the smaller particles in good, loamy soil, an adherence strong enough to resist the downward drag of gravity. If it were not for this property, there would be precious little plant life outside the ocean.

Much more could be said about water, not only because we had to condense the account and leave many things out, but also because, as science advances, more and more amazing properties are discovered. Even though water is relatively simple when compared to living things, it is nevertheless intricately fit for its role as the liquid of life, from the prebiological conditions in regard to the hydrological cycle, temperature moderation and mineral circulation, to its strictly biological use in the bodies of complex living beings.

We dwell on the extraordinary fitness of water at such length because we are trying to break a spell, and for that, just a dash of the wonders of water isn't enough. It has been a great surprise that water, seemingly humble and humdrum, is so wonderful a solvent against the disanthropic materialist reductionism that has had us in its grip for so long. Water proves to be a thing of genius, something both simple but deeply and richly contrived. In this, water is much like the simplest of Shakespeare's ingeniously wrought lines, "To be or not to be," which, although consisting of a minimum of letters (six) and words (six), opens most elegantly onto the nearly immeasurable density of meaning of the entire *Hamlet*. Water, a chemical word consisting of only two elemental letters, contains densely packed within its simple structure all the needed powers that complex and even intelligent life demand from its principal liquid.

And by "principal liquid" we mean principal not only in sustaining life,

---

[24]Our thanks to astrobiologist Guillermo Gonzalez for the point about rainbows. Also, we should not be tricked by children's books depicting mice or miniature people warming themselves by a little fire made by tiny sticks. Smaller vegetation would provide a poor heat source even for small creatures because the smaller the stick, the more quickly it burns, as anyone who has built real fires knows. Also, the kind of heat necessary to smelt metal cannot be had by brush or lesser vegetation. In fact, to smelt well, human beings had to build clay kilns to contain and hence intensify the heat. It would be interesting, in that regard, to find that Earth had an anomalously large percentage of silicon in its upper crust that happily provided the material for such kilns.

but in guiding scientists into the deep mysteries of life. That is why, as we have seen, the presence of liquid water is the first criterion by which astrobiologists judge habitable zones in galaxies and solar systems. We also recall that it was not only water's evident beauty and power that drew philosophers to exalt it as an element, but also its chemical powers that made it omnipresent among the earliest artisans, the later alchemists, and all chemists. In particular, it was water's commonness and chemical simplicity that triggered Antoine Lavoisier's discovery of the elements. And as we have just seen, it is water, perhaps more than any other compound, that reveals the presence of nature's extraordinary prebiological fitness for the life to come. Water is truly meaningful.

All of this because hydrogen and oxygen harmonize—an important point in regard to all of the elements that brings us to another criterion of genius. Imagine the elements as musical notes arranged according to the law of octaves, since as we see from our table (see chapter five), the structure of the order of the periodic table is defined by the eight vertical columns or groups, headed by H, Be, B, C, N, O, F and He. As with a single note, an element can exist independently;[25] but just as the richness of the notes is revealed in chords, so also the real chemical richness of the table is made manifest when elements are composed into complex compounds. Single notes are striking, but to strike single notes or even single chords is not yet a display of the full depth of harmony.

In addition to this "vertical" relationship of notes in chords, there is the "horizontal" relationship of notes and chords in melody. In much the same way, the real depth inherent in the elements is revealed not fully in single compounds, but in the combinations of compounds and the kinds of structures, reactions and activities that these combinations make possible. When a composer so organizes all the aspects of music to create a unified harmonious whole, she has used the entire depth of music that exists in the potentialities of the notes by creating a single piece of music that fits into a genre or species of music. Just so, in nature, the potentialities of the chemical elements are only fully realized when we see them organized into complex, unified, harmonious wholes, and it is there, in living creatures, that we see the entire depth and culmination of chemistry.

Disanthropic reductionism, of course, cringes at such an assertion. But

---

[25]Some, such as those in Groups I and II, are so highly reactive that they don't exist in nature except in compounds, but chemists have contrived ways to isolate and store them.

consider what reductionism means in regard to music, the kind of boorish-
ness it entails. Imagine hearing the following account of one of Wolfgang
Amadeus Mozart's symphonies: "We have been able to prove that this particular symphony is actually reducible to a series of notes that happen to be
played both at the same time in chords and one after another, creating a
string of disturbances in the air caused by different frequencies. We realize,
of course, that these disturbances cause further disturbances in the audience, due in part to the presence of Earth's particular atmosphere and in part
to the effect such disturbances have on the apparatus of the ear as transmit-
ted by neurons to the brain—so disturbing, in fact, that some break into in-
voluntary tears, remarking that they seemed to be hearing the very harmonies of heaven. Happily, we now know that there is nothing more to
Mozart's work in particular and to music in general than mere notes, themselves reducible to waves disturbing air."

Such reductionism displays the kind of bluntness of soul we found in Sigmund Freud, which could reduce the glory of *Hamlet* to the irrational gur-
glings of sexual desire. It is the precise bluntness of soul that led Charles
Darwin to reduce the origin of music to mating calls and, hence, to the sexual desire that drives sexual selection.[26]

Happily, the advance of science itself is undermining such reductionism.
We have seen this clearly in the way that science is, more and more, becoming anthropically defined, pointing upward toward intelligent, organic life
rather than dragging everything down to the levels of quarks and photons.
It is not only that we are finding ever more delicately contrived conditions
that must obtain before there can be complex life; it is also that we are finding extraordinary fine-tuning and design for the sake of biology in the chemical elements and simple compounds themselves.

This knowledge is only possible in light of our extended examination
of living things in a supportive biosphere. Again, we cannot truly know
what carbon is or grasp the genius of its relationship to other elements on
the periodic table until we see its potentialities expressed in living things.
Water by itself cannot reveal its full glories; we could not discover the extraordinary properties of water as the liquid of life without life. But life not
only requires a mind-boggling integration of chemical complexity in indi-

---

[26]See the section on "Voice and Musical Powers" in Charles Darwin's *Descent of Man, and Selection in Relation to Sex* (Princeton: Princeton University Press, 1981), 2:330-37, esp. the sum-
mary on pp. 336-37. We note that Darwin's discussion of music comes in a chapter entitled
"Secondary Sexual Characters [*sic*] of Man."

vidual living things; the most complex life depends on the planet's sophisticated integration of animal with plant life, existing within a supportive biosphere involving distinct prebiological cycles such as the hydrological, carbon and nitrogen cycles. All this in turn depends on intricate fine-tuning at the solar, galactic and cosmological levels, which, taken together, makes the existence of complex life on Earth a culminating symphony of such magnitude and grandeur that it leads us to humble recognition of the genius of nature.

Given that the conditions that allow for this symphony are becoming ever more minutely specified as Earth's particular conditions, we are drawing ever closer to the surprising conclusion that (contra Weinberg) we *do* have a very special relation to the universe. Apparently human life is not just a more-or-less farcical outcome of an accidental chain reaching back to the first three minutes. Instead, it appears we were built in from the beginning.

Yet there's more. We had to wash away the acids of reductionism to see the genius of *Hamlet* and *The Tempest*. We can see that the unity of one of these works—that which brings all of its parts into a harmonious, dramatic whole—is in large measure the effect of Shakespeare's particular mind at work. We should not underestimate the importance of this point. We did not view either Shakespeare or his genius directly, but we recognized genius through one of its effects—a particular play, ingeniously wrought.[27]

And while the recognition of his genius is a humbling one, it is sweetened by the knowledge that we are not without intellectual kinship to Shakespeare. We are able to participate in his genius by viewing and studying his plays; and in this participation, we discover in ourselves the same kinds of intellectual powers that are irreducible to some less noble or even ignoble drive. Wading into the depths of *Hamlet* or *The Tempest* without the blinders and chains of reductionism is to discover depths in ourselves, depths that are not susceptible to explanation in Freudian, Marxist or even Darwinian terms. For the same reason, to plumb the depths of Euclid's work makes us understand not only how ingenious this greatest of geometers was, but how extraordinary and peculiar we rational animals are.

And the chemical order in nature, so beautifully distilled in the periodic table of elements—a drama stretching from the big bang all the way

---

[27] And here we need not engage the far-fetched claims that Shakespeare was really someone else because the relevant point is that the play's creative genius is understood to be a *person*.

to our discovery of the big bang—is a drama of high genius, and one that seems meant to be unraveled by human genius. Just as *The Tempest* or *Hamlet* lies outside the creative reach not only of madly typing monkeys, but also of all the madly typing human fans of the Bard, so too the assembly of the periodic table was beyond the reach of all but a few great scientists, those possessing intellectual powers inexplicable according to the reductionist tenets of materialism in general and hopelessly beyond any connection to natural selection in particular. Such genius depends for its existence on an irreducibly complex series of conditions stretching from the universe's origins, through the synthesis of the elements in the periodic table, and right up to the peculiar conditions defining the delicate biocentric balance of Earth. And in this series—as we have seen with carbon and water—these potentialities existed long before life and, therefore, before Darwinian natural selection could serve as a possible creative force.

The intrinsic order of the table makes all this possible, and it is written into nature itself, not in a hopelessly obscure script but in a way that we can decipher. Nature and intelligibility are carefully entwined, a work of art for rational animals, a laboratory for the invention of both actors and audience.

The Dictionary of the History of Ideas discusses the term *genius* at length. One portion is particularly apropos:

> The Renaissance concept of the *divino artista* ("the divine artist") had a double root. On the one hand, it was derived from Plato's theory of the *furores,* the inspired madness of which seers and poets are possessed; on the other hand, it looked back to the medieval idea of God the Father as artist, as architect of the universe. . . . As early as 1436, Leon Battista Alberti suggested in his treatise *On Painting* that the artist may well consider himself, as it were, another god, an *alter deus.* . . . Shaftesbury, who according to Ernst Cassirer (1932; 1955) rescued the term "genius" "from the confusion and ambiguity that had previously attached to it," goes on to characterize the inspired poet, the real Master, as "a second Maker; a just Prometheus under Jove."

Given that nature conforms far more profoundly to the qualities associated with the works of genius than any human masterpiece, the student of nature may say with some confidence that the scientific news of the Author's death has been greatly exaggerated. *The Dictionary of the History of Ideas* includes another detail equally relevant to this discussion, noting, "Although we are reminded that the man of the second half of the twentieth century no longer believes in geniuses," they can hardly be abolished by an act of "cultural will."[28]

The materialist who attempts to abolish genius is cutting off his nose to spite his face—an unwise act especially for a scientist, who is sure one day to need his sense of smell.

---

[28]"Genius," in *The Dictionary of the History of Ideas: Studies of Selected Pivotal Ideas*, ed. Philip P. Wiener (New York: Charles Scribner's Sons, 1973-1974), accessed online at <http://etext.lib.virginia.edu/cgi-local/DHI/dhi.cgi?id=dv2-36>. The work also discusses Immanuel Kant's theory of genius in the entry "Genius from the Renaissance to 1770." There the dictionary paraphrases Kant to the effect that "Genius, freedom, and living organisms are elements which cannot be explained mechanically (Tonelli, 1966)." Already by Kant's time the materialist spirit was at work challenging this idea and, with the Darwinian framework in place, the wisdom of Kant and of Western culture on this point had been rejected.

8

THE REEMERGENCE OF
THE LIVING CELL

Late twentieth-century genetics has met with two failures,
both of them a result of its reductionism and its predilection for artificiality.

Giuseppe Sermonti, *Why Is a Fly Not a Horse?*

MOVING FROM CHEMISTRY TO BIOLOGY, let's now focus not on a human text, but on the disappearance of a living text, one obscured for the same reasons the texts of William Shakespeare were obscured—materialist reductionism. Recall what Sigmund Freud did to *Hamlet:* he took this grand and timeless expression of the profundities of familial love, friendship, romance, revenge, political machinations, divine justice and the afterlife and reduced it to the crudest, underlying material causality, the incestuous and relentless id of Hamlet.

On this view, the drama isn't real; it is merely an appearance, one of a number of possible epiphenomena aimlessly generated by a bubbling subconscious cauldron of sexual drives. Freud reduced literary criticism to psychology, and psychology to sexual instinct. And it is in keeping with his materialism to further reduce even these sexual urges to chemistry and, ultimately, physics. On Freud's view, the tragedy, as it presents itself, is no more real than the drama of our everyday existence—our complex reasons for acting, our search for truth and beauty, our loves and hates, our intricate familial relationships, our art, politics and literature—are all reducible to primitive, brutish urges.

This reductionist flattening of Shakespeare's drama was rooted in the larger reductionist-materialist worldview Freud inherited and embraced, a view that grinds the intricate fabric of life, all life, to a lifeless, homogeneous pulp. At the heart of this view is the belief that nothing essential is lost in the grinding because there is no ultimate difference between living and non-living things. This assumption leads to a poisonous paradox: a fundamental aim of modern biology, life science, is to eliminate the commonsense notion that "living" is a fundamental category worthy of its own science.

How did this occur? The reasons are complex, but the Victorian age played a critical role. Charles Darwin unleashed an explanatory mechanism on the world of biology, the idea that natural selection working on random variation allowed the first organism to evolve into the diversity of life we find around us. Darwin may have intended for his theory to leave the organism intact, but like the Frankenstein monsters of science fiction, it didn't. Biologist Brian Goodwin explains:

> A striking paradox that has emerged from Darwin's way of approaching biological questions is that organisms, which he took to be primary examples of living nature, have faded away to the point where they no longer exist as fundamental and irreducible units of life. Organisms have been replaced by genes and their products as the basic elements of biological reality. This may seem to fly in the face of all common sense, but stranger things have happened in the name of science.[1]

Against this reductionism, Goodwin takes the commonsensical position that "organisms are as real, as fundamental, and as irreducible as the molecules out of which they are made. They are a distinct level of emergent biological order, and the one to which we most immediately relate."[2] In short, Goodwin is making the astounding assertion that living things are real. He is arguing that dogs, cats, katydids, crawdads, blue-bottle flies, cows and even human beings—all the creatures we meet in the everyday drama of existence—are not essentially urges or genes or atoms, but essentially dogs, cats, katydids, crawdads, blue-bottle flies, cows and even human beings.

Now it should strike some as odd that Goodwin is merely affirming what any child, what any farmer, would find obvious. Imagine if this eminent biologist burst into the bustle of an ordinary farm and, with the breathless exhilaration of a revolutionary, announced to the busy farmer, "This cow is *real!*" The farmer would justly wonder how biology could become so disconnected from the obvious that the reality of Henrietta the Holstein should

---

[1]Brian Goodwin, *How the Leopard Changed Its Spots: The Evolution of Complexity* (New York: Charles Scribner's Sons, 1994), p. vii.
[2]Ibid, p. x.

have been a matter of doubt.

Now to Goodwin's credit, he is not attempting an intellectual revolution among farmers, who have not lost their common sense, but among his fellow biologists, who have. So it is that Goodwin and fellow biologist Gerry Webster are forced to trumpet the obvious as if it were a matter of controversy for, as they lament, "the organism as a real entity, existing in its own right, has virtually no place in contemporary biological theory."[3] Of course, for reductionists in biology, that is not bad news, for they view as a sign of progress the reduction of phenotype to genotype, organism to DNA, form to matter, biology to chemistry. But for Goodwin, Webster and others of the new structuralist or emergentist school in biology, the disappearance of the organism represents an unwarranted rejection of the proper and real object of the science of biology.

Nor is the reductionism confined to the macroscopic world. It is applied all the more assiduously to the microscopic world, where the commonsense notion of a living cell has been largely rejected by biochemists and molecular biologists. In the words of biochemist Franklin Harold, "As a subject for serious inquiry, the category of 'life' has all but vanished from the scientific literature; it is the particulars of life, not its nature, that fill the numberless pages of scientific journals."[4] Harold continues his complaint:

> During the past half-century, the program of analyzing the structure of cells in search of their common denominators has been pursued to the molecular level with notable success and with mounting zeal. This single-minded concentration on the relatively tractable problems of chemical structures and interactions has been accompanied by neglect of the higher levels of biological order, often to the point of absurdity. Surely, one plain lesson to be read in the prevalence of cellular architecture [in all living things] is that organized complexity is one of the essential characteristics of life. From this viewpoint, the significance of cells is that they represent the minimal level of organization capable of displaying the activities we associate with life, including self-reproduction. . . . Cells can only be understood as organized systems, and . . . the physiological and molecular levels of inquiry are necessarily complementary. . . . That this self-evident proposition should remain in practice so much of a minority view never ceases to astonish me.[5]

---

[3]G. Webster and B. C. Goodwin, "The Origin of Species: A Structuralist Approach," *Journal of Social and Biological Structures* 5, no. 1 (1982): 16.

[4]Franklin Harold, *The Way of the Cell: Molecules, Organisms and the Order of Life* (Oxford: Oxford University Press, 2001), p. 9.

[5]Ibid., p. 31.

Of course, Harold believes the study of chemical substructure has its place. It just shouldn't level the place. So while admitting that "one cannot reflect usefully on the phenomenon of life without taking account of its material basis," he takes issue with a single-minded reductionism that would gravely distort what it ultimately attempts to explain—the living, acting cell.[6] This distortion, Harold argues, is nowhere more evident than in biochemists' normal laboratory procedure:

> When [we] biochemists set out to tackle a problem, our first step is commonly to grind the intricate fabric of cells and tissues into a pulp (a homogenate, as we say in the trade). This is a significant act, representing a drastic reduction in the level of organization. It allows us to treat living matter as a mixture of chemicals. . . . To be sure, something is sacrificed by this violent procedure— not only life itself, but all the spatial order that impresses anyone who inspects a photomicrograph [of a cell]. But never mind: biochemists still cherish the premise that nothing irretrievable is lost by homogenization, and that given the macromolecules, all the essentials are present and accounted for. We know quite well that this cannot be true, but the focus on the molecules defines that layer of knowledge that we designate as biochemistry or molecular biology, and undergirds our professional identity.[7]

We see, then, that both on the microscopic level of the cell, the simplest level of life, and on the macroscopic level of the larger, far more complex multicellular organisms, the reality of the living thing has been lost, destroyed by the acids of reductionism in an attempt to explain all living wholes as mere accumulations of material parts and to reduce biology to chemistry.

But, of course, these complaints, however deeply felt, might be entirely unjustified. *If* in fact such reductionism is correct in its approach and hence in its results, then we would merely be complaining about especially acute intellectual growing pains: living things can be reduced to their parts, and they aren't substantively real, so now that we know the truth, we'll just have to buck up and get used to it. In order for us to have a justified complaint, materialist reductionism must be shown to be fundamentally erroneous.

There are two important ways to show that some belief or theory about nature is bogus. We may dispute its fundamental assumptions through the power of pure philosophy, so to speak. The problem with philosophy, un-

---

[6]Ibid., p. 34.
[7]Ibid., p. 35.

fortunately, is that it is a human thing, and so there is plenty of room for error and disagreement and even more room to hide in entirely abstract houses of our own construction. The early Greek philosophers quarreled over what the fundamental substance or substances in nature were. One persuasive case butted against another persuasive case. It was all quite logical but never especially empirical, for nature had not yet had sufficient say. Another way to root out error is simply to let an error run full throttle and look for nature to contradict it. The alchemists' passionate search for the philosopher's stone was the best way to find that it doesn't exist, just as the best way to destroy the theory of phlogiston was to give the phlogistians full rein to track it down. And so it is with the reductionist belief that complex life is essentially a lucky aggregation of chemicals, that the cell can be produced by an accretion of chemical parts, and that this in fact occurred on the early Earth. It is precisely here that the dogma of reductionism, however loudly it barks, hits the end of its rope.

In *The Origin of Species,* Darwin tactfully avoided any discussion of how it could have been that the first living things might have arisen, and there were good reasons for his doing so. Most importantly, the main engine of evolution, natural selection, requires the existence of living things, for only living things can self-replicate and pass on traits. But there were other reasons as well. As Adrian Desmond and James Moore have made clear in *Darwin: The Life of a Tormented Evolutionist,* Darwin was keenly aware of the atheistic implications of his theory, and he went to great pains to avoid the charge of atheism. Sidestepping the question of how the first living things arrived allowed Darwin to provide a convenient and conciliatory wisp of theism in the finale of *Origin*:

> There is a grandeur in this view of life, with its several powers, having been originally breathed by the Creator into a few forms or into one; and that, whilst this planet has gone cycling on according to the fixed law of gravity, from so simple a beginning endless forms most beautiful and most wonderful have been, and are being evolved.[8]

Privately, however, Darwin hoped that the breath of God would be replaced by simple chemistry (based on the typical nineteenth-century assumption that the chemical foundation of life *was* simple). As he said in his now famous private letter, "If (and oh! what a big if!) we could conceive in some

---

[8]Charles Darwin, *The Origin of Species*, 6th ed. (New York: Mentor, 1958), p. 459.

warm little pond, with all sorts of ammonia and phosphoric salts, light, heat, electricity, &c., that a proteine compound was chemically formed ready to undergo still more complex changes,"[9] then perhaps the need for a Creator of life could be circumvented and nonliving chemicals could bring forth life.

But we must understand that this "big if!" arose from his commitment to a particular philosophy, not to some astounding evidence. This is all the more clear when we realize that the more general theory of evolution itself was not original to Darwin, but was inherited from and entailed by the larger materialist framework Darwin accepted. As discussed earlier, the notion of evolution itself is not even modern but can be traced to ancient Epicurean materialism.[10] Epicureanism defined the universe anti-theistically: according to Epicurus, and even more clearly, to his Roman disciple Lucretius, the world was *not* created or designed by the gods; rather, the cosmos and all within it arose from the purposeless, random banging together of atoms.[11] This is a very modern-sounding position, but one that was based on conjecture—or more accurately, based on an explicit desire to rid the universe of a Divine Cause. And nineteen centuries before Darwin, we also find in Lucretius the original warm little pond. In book five of his *De Rerum Natura*, we are taken on an entirely imaginative and poetic journey back to the "childhood of the world."[12] Of course, that means that, since no divinity was responsible for creating living things, nonliving atoms will somehow have to bring forth life on their own. Lucretius opts for cooking the atoms on Earth "with the aid of showers and the sun's genial warmth." He begins with a kind of omniscient narrator's description of the early Earth:

> [There] was a great superfluity of heat and moisture in the soil. So, wherever a suitable spot occurred, there grew up wombs, clinging to the earth by roots.

---

[9]Francis Darwin, *The Life and Letters of Charles Darwin*, 3 vols. (New York: Johnson Reprint Corporation, 1969) vol. 3, p. 18. Darwin's rumination is reported by his son Francis in a footnote to a letter to J. D. Hooker, written on March 29, 1863. The rumination itself was written in 1871, and Darwin's larger point is that such prebiological compounds must still be produced today but fail to yield production of new creatures because the compounds would "at the present day . . . be instantly devoured or absorbed, which would not have been the case before living creatures were formed." In other words, Darwin thought that he had to explain why the Earth didn't continue to churn out new evolutionary lines from warm little ponds.

[10]Again, see Benjamin Wiker, *Moral Darwinism: How We Became Hedonists* (Downers Grove, Ill.: InterVarsity Press, 2002), esp. chaps. 2, 4 and 8.

[11]Lucretius *De Rerum Natura (On the Nature of Things)* 5.419-31. See also 1.1021-51, 2.167-81, 2.1058-1104, 4.823-57, 5.76-90, 5.198-234, 5.417-31.

[12]Ibid., 5.780. All quotations are from Lucretius *On the Nature of Things*, trans. R. E. Latham, rev. John Godwin (New York: Penguin, 1994).

These, when the time was ripe, were burst open by the maturation of the em-bryos, rejecting moisture now and struggling for air. Then nature directed to-wards that spot the pores of the earth, making it open its veins and exude a juice resembling milk, just as nowadays every female when she has given birth is filled with sweet milk. . . . Here then is further proof that the name of mother has rightly been bestowed on the earth.[13]

Lucretius was advocating a form of spontaneous generation, but not one in which living creatures magically or miraculously sprang directly from the mud. For Epicurean materialists, as for their modern intellectual heirs, the distinction between living and nonliving was merely a mental construct—often useful, but not real the way a stone is real. Therefore living wholes were caused by, and could therefore be reduced to, a fortuitous arrange-ment of atomic parts.

What should strike us about Lucretius's account is that he was hardly driven to it by powerful evidence. How could he, given the state of science in the first century before Christ? Clearly, the conjecture is based on the dictates of his materialism. He sought an evidentiary hook to hang his theory on only af-ter he had imagined it: when wood and earth rot into a murky substance, he tells us, the atomic structures disintegrate, so that "when they are fairly well rotted by showers," they "give birth to little worms, *because* the particles of matter are jolted out of their old arrangements by a new factor and combined in a way that animate objects must result."[14] And so it is that "everything de-pends on the size and shape of the . . . atoms and on their appropriate mo-tions, arrangements and positions."[15] Worms can be found in dirty, rotten wood; therefore, the materialist cosmos is all there is, ever was or ever will be. On close inspection, the argument itself resembles rotten wood.

If we disregard Lucretius's poetic embellishments, his account is not far removed, in its principles and premises, from Darwin's. Neither invoked magic or miracle or intelligent creation. Darwin's own cherished dictum was *natura non facit saltum,* "nature does not make a leap." His confidence against such leaping was based, in part, on his unjustified belief that getting to the first and simplest organism wasn't much of a leap—a leap of materi-

---

[13]Ibid., 5.799-823.
[14]Ibid., 2.893-900 (emphasis added).
[15]Ibid., 2.892. We leave out the word *sentient* so as not to be confusing. By "sentient atoms," Lucretius meant the inanimate atoms so arranged as to produce the purely material phenom-enon of sensation, not that there were peculiar atoms capable of sensation. As he argues just above this section, "Whatever is seen to be sentient is nevertheless composed of atoms that are insentient" (2.865).

alist faith on his part, rather than one based on science.

Those eager to expunge God's fingerprints from nature weren't concerned by this shortcoming in Darwin's material explanation for life, because Darwin and his contemporaries thought a single cell was a simple blob of protoplasm. How hard could it be for nature to randomly produce something so simple?

At least for the time being, ignorance was theoretical bliss. In Darwin's time, and for some time after, the cell was a black box, a mystery, hardly a firm foundation on which to build such grand assumptions. But in the twentieth century, scientists were able to open that black box and peek inside. There they found not a simple blob but a world of complex circuits, miniaturized motors and digital code. We now know that even the simplest functional cell is almost unfathomably complex, containing at least 250 genes and their corresponding proteins, each one extraordinarily difficult to produce randomly and none of which can function apart from the intricate structure of the cell. Biophysicist Dean Kenyon was a leader in the effort to offer a materialist explanation for the origin of cellular life, coauthoring *Biochemical Predestination,* a leading monograph in the 1960s about how life could have emerged by purely undirected processes.[16] But Kenyon later concluded that the regularities of chemistry couldn't produce the information-rich structures essential to even the simplest life.

If such an intricate little world, capable of self-replication, suddenly leaped into being, one would rightly be suspicious that mere chemistry was not at work. Therefore, origin-of-life researchers still wedded to the larger scheme of materialism have focused on the slow generation of significant chemical parts of the cell in hopes that, if those parts could somehow arise, then somehow the rest of the parts would follow and the whole cell would eventually fall into place. The problem facing such efforts is, again, the *integrated* nature of the living cell: it depends on a multitude of parts for its existence; these parts are interdependent, so that if one did happen to arise *(per impossibile),* it would stand as idle and functionless as a carburetor in a pile of scrap metal. It's not just a matter of getting all the necessary individual parts in one location—in itself, a potentially insurmountable difficulty—but, even more, a matter of explaining how they became functionally integrated according to blueprints that do not yet exist.

How to get around these difficulties? Undoubtedly the most famous way

---

[16]Dean Kenyon and G. Steinman, *Biochemical Predestination* (New York: McGraw-Hill, 1969).

was the positing of the existence of a prebiotic chemical soup in which—because of its chemical content, concentration and felicitous conditions—the parts could be simmered into living wholes (or at least prepare the way for the wholes-to-be). Such assumptions formed the theoretical background of the famous Miller-Urey experiment in the mid-twentieth century, wherein an electric current (simulating a lightning storm) was passed through a mixture of gaseous methane, ammonia, hydrogen and water—the gases assumed to be found in Earth's earliest atmosphere—resulting in the formation of some carbon-based organic compounds. This piece of laboratory magic seemed to give chemical flesh to the Epicurean-Lucretian-Darwinian idea of spontaneous production of Earth's first life. Only now, in between the mythical life-giving mud and the worms, there was a long string of "atoms and . . . their appropriate motions, arrangements and positions"—methane, water vapor, ammonia, amino acids, nucleic acids, RNA, DNA, lipids, and so on up to the first cell. From here, the path was already laid out: the "appropriate motions, arrangements and positions" of the DNA via millions of years of Darwinian evolution would finish the work of turning mud to worm.[17]

In showing the ancient intellectual roots of the warm little pond, we are not simply asserting that either Darwin or Stanley Miller and Harold Urey had Lucretius in hand as they ruminated on the necessities of the production of life from nonliving chemical broth. As for Darwin, Lucretius's poem had been reintroduced to the West in the fifteenth century and by Darwin's time was already an established classic.[18] The idea could have come from a number of sources, not the least of which was Darwin's own grandfather Erasmus Darwin, whose poem "The Temple of Nature" includes the following Lucretianesque rumination:

Then, whilst the sea at their coeval birth,
Surge over surge, involved the shoreless earth;
Nursed by warm sun-beams in primeval caves
Organic life began beneath the waves. . . .
Hence without parent by spontaneous birth,
Rise the first specks of animated earth.[19]

---

[17]Lucretius *De Rerum Natura* 2.892.
[18]Lucretius's work was an "established classic" both in the sense of being widely read and in the sense of having the status of a standard of Latin literature to be labored over by students.
[19]Erasmus Darwin, "The Temple of Nature," quoted in John Farely, *The Spontaneous Generation Controversy from Descartes to Oparin* (Baltimore: John Hopkins University Press, 1977), pp. 44-45.

Publicly, at least, Darwin remained aloof from the question of the origin of life. In his private ruminations, as we have seen, he did entertain notions of abiogenesis and even seemed to accept the notion of "living atoms" quite popular among the radical intellectual set in the early nineteenth century, wherein atoms had a kind of "organizing energy" of their own, thus providing for a kind of materialist creative energy facilitating the organization of living things from the chemical elements.[20]

As for Miller and Urey, there were more proximate if equally telling intellectual and philosophical influences: Alexander Ivanovich Oparin (1894-1980) and J. B. S. Haldane (1892-1964).

Nobel laureate Christian de Duve's remarks about Oparin and Haldane—somewhat offhand but still revealing—are worth quoting. Noting that Oparin and Haldane were both "confirmed Marxists" and "militant defenders of dialectical materialism," de Duve muses that

> one may wonder to what extent ideology had something to do with their desire to explain life as a naturally emerging phenomenon in the evolution of the Earth. It certainly had with Oparin, whose book is peppered with references to the philosopher Friedrich Engels. Rumor even has it that he was set on the problem by the Party.[21]

Oparin hardly needed such encouragement. He was convinced that Karl Marx's intellectual partner Friedrich Engels had properly foreseen the necessity of the production of living cells from nonliving chemicals in Engels's *Dialectics of Nature*. There Darwinism is taken to be the intellectual culmination of modern materialism as manifested in biology,[22] although Engels, of

---

[20]On Darwin's remaining aloof from the controversies regarding spontaneous generation, see Farely, *Spontaneous Generation Controversy*, p. 127. On the notion of living atoms, see Adrian Desmond and James Moore, *Darwin* (New York: W. W. Norton, 1994), p. 223.

[21]Christian de Duve, *Blueprint for a Cell: the Nature and Origin of Life* (Burlington, N.C.: Neil Patterson, 1991), p. 109. Perhaps de Duve is being cynical, although it is quite enlightening to read about Ivanovich Oparin's sycophantic escapades with the Communist Party. A must-read on the subject is Robert Shapiro's delightful account in *Origins: A Skeptic's Guide to the Creation of Life on Earth* (New York: Summit, 1986), pp. 140-54. For more detail, see David Joravsky, *The Lysenko Affair* (Cambridge, Mass.: Harvard University Press, 1970). J. B. S. Haldane was apparently less doctrinaire than Oparin, even denying that he was, strictly speaking, a pure materialist. See his essay "Some Consequences of Materialism," in J. B. S. Haldane, *Science and Human Life* (New York: Harper & Brothers, 1933), pp. 155-69. Here, he offers his assent for a kind of Hegelianism, based in part on the problem that thought causes for strict materialism (pp. 155-56, 167-68).

[22]See the introduction to Friedrich Engels, *Dialectics of Nature* (1927; Moscow: Progress Publishers, 1972), pp. 20-31. For his strange attempt to give his account a Marxist spin, read pp. 306-8.

course, gives it a Marxist interpretation according to the canons of dialectical materialism, where the survival of the fittest is recast in terms of the great and fruitful upward march of opposites (i.e., the Marxian dialectic of thesis, antithesis and synthesis). Engels writes:

> Dialectics . . . prevails throughout nature, and so-called subjective dialectics, dialectical thought, is only the reflection of the motion through opposites which asserts itself everywhere in nature, and which by the continual conflict of the opposites and their final passage into one another, or into higher forms, determines the life of nature. . . . All chemical processes reduce themselves to processes of chemical attraction and repulsion. Finally, in organic life the formation of the cell nucleus is likewise to be regarded as a polarization of the living protein material, and from the simple cell onwards the theory of evolution demonstrates how each advance up to the most complicated plant on the one side, and up to man on the other, is effected by the continual conflict between heredity and adaptation.[23]

As might be expected—since Marxists believe everything in regard to human historical development depends on the effects of human labor—even the transformation from ape to human occurred according to the dictates of Marxian dialectical materialism, as Engels makes clear in a chapter entitled "The Part Played by Labour in the Transition from Ape to Man."[24] In regard to the transformation of chemicals to the first cell, Engels sketches the following scenario, summing up his account of cosmic evolution with the origin of life on Earth:

> If, finally, the temperature becomes so far equalized that over a considerable portion of the surface [of the Earth] at least it no longer exceeds the limits within which protein is capable of life, then, if other chemical pre-conditions are favorable, living protoplasm is formed. What these pre-conditions are, we do not yet know, which is not to be wondered at since so far not even the chemical formula of protein has been established—we do not even know how many chemically different protein bodies there are—and since it is only about ten years ago that the fact became known that completely structureless protein exercises all the essential functions of life: digestion, excretion, movement, contraction, reaction to stimuli, and reproduction.
>
> Thousands of years may have passed before the conditions arose in which the next advance could take place and this shapeless protein [could] produce the first cell by formation of nucleus and cell membrane.[25]

---

[23]Ibid., p. 211.
[24]Ibid., pp. 170-83.

"Completely structureless protein"? One is struck at the contrast between how little Engels actually knows about biology (based, as he admits, on how little his contemporary biologists knew about the cell in general and the powers of protein in particular) and his supreme confidence in his overall sketch.[26] The unfinished manuscript of the *Dialectics* ends with a telling comment by Engels to himself: "(All this has to be thoroughly revised)."[27]

Engels himself would not undertake such a revision, but Oparin would. In his now famous *The Origin of Life* (1938), Oparin set himself the task of giving chemical flesh to the Marxian philosophic framework of dialectical materialism (a work that was initiated with his paper on the subject in 1924).[28] Interestingly, Oparin's historical account of the philosophic and scientific speculation about the origin of life culminates in the arguments not of a scientist, but of the philosopher Engels, and it shows Oparin's intellectual indebtedness to materialism in general and Marxism in particular: "Engels shows that a consistent materialistic philosophy can follow only a single path in the attempt to solve the problem of life. Life has neither arisen spontaneously nor has it existed eternally. It must have, therefore, resulted from a long evolution of matter."[29]

It is important to note that, in arguing that life had not arisen "spontaneously," Oparin meant to reaffirm Darwin's assumption, *natura non facit saltum,* in regard to the origin of life: there couldn't be a leap into complexity; rather, the rise of the living from the nonliving had to be the result of "a long evolution of matter." (Also, being a Marxist, Oparin held that "mere" materialism was an insufficient engine of transformation; rather, the change had to be explained according to *dialectical* materialist canons of opposition and transformation.)[30] Oparin's adherence to Marxism became quite pronounced, especially after Joseph Stalin's power began to dominate even the pursuit of science in the Soviet Union. In the opening paragraph of his *Or-*

---

[25]Ibid., p. 33.

[26]From his chapter on biology, it is clear that Engels was relying, in great part, on the speculations of the Darwinist Ernst Haeckel. See esp. ibid., pp. 303-6.

[27]Ibid., p. 311. Since this book was a collection of essays published posthumously, he may not have intended to publish them in the state they were in. Even so, the unfinished state is quite revealing.

[28]A. I. Oparin, *The Origin of Life,* trans. Sergius Morgulis. 2nd ed. (New York: Dover, 1953.) The Dover edition is an unabridged republication of the original English translation done in 1938.

[29]Ibid., p. 33. Note Oparin's praise of Engels's protein theory as prescient (p. 136).

[30]On the importance of dialectical materialism for the details of Oparin's theory, see Farely. *Spontaneous Generation Controversy,* chap. 9.

igin of Life, Oparin placed the origins quest firmly within a Marxist framework. "The question of the origin of life, of its first appearance on Earth, still occupies the human mind, as it has done since the most remote antiquity," remarked Oparin. "During different epochs and at different stages of civilization the question was answered differently, but this question was always the focal point of a sharp philosophical struggle which reflected the underlying struggle of social classes."[31] He seems to be suggesting that he had to get teleology out of the origin of life in order to get ruling-class religion off the back of the worker. A more dogmatic approach to the origin-of-life question would be hard to imagine.

Haldane was actually the one who coined the term "hot dilute soup" to describe the "vast variety of organic substances . . . including sugars and apparently some of the materials from which proteins are built up" that "must have accumulated" in "the primitive oceans."[32] But it was Oparin who first attempted to spell out in detail the existence and ingredients of the broth in his *Origin of Life.*[33] This attempt to explain the origin of life within a Marxist-materialist framework came to be called the Oparin-Haldane hypothesis, and it was this hypothesis (as historians of science agree) that helped form the general intellectual parameters of the Miller-Urey experiment to recreate the "hot dilute soup" in the laboratory.

None of this is to say that prebiotic soup is merely a Communist idea and hence suspect. Rather, Marxism itself is just one of the many "species" of modern materialism, and true to its genealogy as a form of reductionism, it *assumes* that life must have arisen by a chance concatenation of chemicals. This brief history merely illustrates that the origin and rise to prominence of the notion that life emerged spontaneously from a primordial chemical soup did not itself have a scientific, evidential basis; rather, it was made necessary by the materialist assumption that life *must* have arisen from some fortuitous chemical mixture. As Oparin put it, "[A] consistent materialistic philosophy can follow only a single path in the attempt to solve the problem of life." Oparin was certainly correct here, for if life did not arise through mere chemical evolution, materialism would obviously be false. Since, for Oparin, materialism could not be false, then the soup must have existed.

It is this wider, modern materialist conviction, flowing toward Miller and

---

[31]Oparin, *Origin of Life,* p. 1.
[32]See Haldane's essay in the collection J. B. S. Haldane, *Science and Human Life* (New York: Harper & Brothers, 1933), pp. 142-54. The famous quote is on p. 149.
[33]Oparin, *Origin of Life,* esp. pp. 106-7, 123-26, 248.

Urey through the more restricted channels of Oparin and Haldane, that formed the intellectual quest to reproduce the legendary prebiotic soup in the laboratory. Now Marxist materialism has not fared well on the stage of history, but if the conjecture about life emerging from a prebiotic soup could be grounded in experimental science, what would it matter where the idea originally came from? It could have been concocted by a society dedicated to enshrining Lawrence Welk as the king of big band jazz, and it might still be true. Good ideas come from compromised sources all the time. The important question is this: Did experimental evidence eventually arrive to rescue the hypothesis?

The prebiotic soup the materialists need must be a very specific kind of alphabet soup, one capable of producing the information-rich code necessary for even the simplest self-reproducing organism, and here is where their assumption bumps up against uncooperative data: the weight of the evidence is strongly against the existence of any such soup. As Hubert Yockey notes in the *Journal of Theoretical Biology*, "If one looks at the geological record, one finds no evidence that a primeval soup ever existed."[34] The most likely reason that we find no evidence of a prebiotic soup, Yockey continues, is quite simple: "There never was a primeval soup."[35]

To understand how the mythic soup could become a fact without evidence, we must return to the classic Miller-Urey experiment that, more than any other event, enshrined the idea of a fortuitous prebiotic soup in both the scientific and popular culture. As has now been generally accepted, the famous Miller-Urey experiment in 1952 that produced some amino acids and other prebiotic building blocks was built on a faulty assumption—the assumption that the conditions necessary to produce the amino acids in the laboratory *must* have existed on the early Earth. Miller and Urey assumed

---

[34]Hubert P. Yockey, "Comments on 'Let There Be Life: Thermodynamic Reflections on Biogenesis and Evolution'" by Avshalom C. Elitzur," *Journal of Theoretical Biology* 176 (1995): 351. The lack of geological evidence is as follows. The production of amino acids in origin-of-life experiments (such as the famous Miller-Urey experiment) entails a far greater production of an "insoluble tarry mixture"—the greater the production of amino acids, the greater the production of the tarry substance. But this tarry substance would have "precipitated out of the primeval ocean and [hence would] have been found in the kerogen of sedimentary rocks" as evidenced by the effects on the amount of C12. Yockey also points out that even those who have held stubbornly to the existence of such prebiotic soup, such as Stanley Miller and Carl Sagan, have admitted that, rather than being rich in prebiological material, it would have been extraordinarily dilute—too dilute (apparently) to leave geological evidence. But if so, it would have been far too dilute for mere chance to have any effect.
[35]Ibid., p. 351.

that the early atmosphere was reducing (i.e., contained no free oxygen), consisting of a hydrogen-rich atmosphere, with the richness of the hydrogen (from ammonia, hydrogen gas, methane and water vapor) providing the proper conditions for the build-up of significant prebiotic molecules.[36]

But if experiments are designed according to the actual atmosphere that scientists now believe the early Earth had—not happily stacked with rich sources of hydrogen, but instead neutral or only slightly (rather than strongly) reducing, composed of carbon dioxide, nitrogen and water vapor—things go badly for the mythic soup. As Noam Lahav notes, even if we allow for "slightly reducing conditions, the Miller-Urey action does not produce amino acids, nor does it produce the chemicals that may serve as the predecessors of other important biopolymer building blocks." He concludes thus:

> By challenging the assumption of a reducing atmosphere, we challenge the very existence of the "prebiotic soup," with its richness of biologically important organic compounds. Moreover, so far, no geochemical evidence for the existence of a prebiotic soup has been published. Indeed, a number of scientists have challenged the prebiotic soup concept, noting that even if it existed, the concentration of organic building blocks in it would have been too small to be meaningful for prebiotic evolution.[37]

Apparently, it was L. G. Sillén who, in 1965, only seven years after the Miller-Urey experiment, first called the soup a "myth" based on its unrealistic chemical assumptions.[38] As Stephen Mojzsis, Ramanarayanan Krishnamurthy and Gustaf Arrhenius have argued, the mythic prebiotic soup was chal-

---

[36]Ibid. It is perhaps important to note as well that Miller and Urey made a reasonable assumption that the early Earth couldn't have an oxidizing atmosphere, since the main source of oxygen is from photosynthesis. But that leaves two other choices—neutral or reducing. They chose the latter because the former would not yield the desired results, and they modeled the atmospheric contents accordingly. Some justification for loading the atmosphere with hydrogen arose from the abundance of the element hydrogen in the universe, but as has been pointed out, such hydrogen, having very low mass, would not have been retained in the atmosphere for very long. For more detailed but readable accounts of the Miller-Urey experiment and its defects, see Fazale Rana and Hugh Ross, *Origins of Life* (Colorado Springs: Nav-Press, 2004), chap. 7; Christopher Wills and Jeffrey Bada, *The Spark of Life: Darwin and the Primeval Soup* (Cambridge, Mass.: Perseus, 2000), chaps. 2-4; and Shapiro, *Origins,* chap. 4. A seminal critique is Charles B. Thaxton, Walter Bradley and Roger Olsen, *The Mystery of Life's Origin: Reassessing Current Theories* (New York: Philosophical Library, 1984).

[37]Noam Lahav, *Biogenesis: Theories of Life's Origins* (Oxford: Oxford University Press, 1999), pp. 138-39.

[38]L. G. Sillén, "Oxidation State of Earth's Ocean and Atmosphere," *Archiv für Kemi* 24 (1965): 431-56, as reported by H. P. Yockey, *Information Theory and Molecular Biology* (Cambridge: Cambridge University Press, 1992), p. 235.

lenged from the time it was publicly enshrined, and rightly so, making even more suspicious its early and sustained reception. In their words, "The idea of such a 'soup' containing all desired organic molecules in concentrated form in the ocean has been a misleading concept against which objections were raised early. . . . Nonetheless, it still appears in popular presentations perhaps partly because of its gustatory associations."[39]

The soup's problems are only beginning. Not only do scientists now doubt that it ever existed, but even if it had, it couldn't have gotten the job done. As Charles Thaxton, Walter Bradley and Roger Olsen demonstrated in their seminal critique of the Miller-Urey experiment, even if Miller and Urey's prebiotic soup and atmosphere had existed on the early Earth, it couldn't have produced amino acids without the same sort of meticulous, intelligent manipulation that Miller and Urey injected into their laboratory experiment.[40]

Then there's the whole matter of getting crude amino acids to assemble into the wildly complex sentences required for even the simplest self-repro- ducing organism. The problem is the same one that plagued natural philos- ophers before the birth of science: grand claims (in this case, for an undi- rected origin of life) are not being properly tested against reality. After nearly a century of research—in which our understanding of the early Earth, of chemistry and of biological complexity has grown exponentially—nature has spoken: the famous soup hypothesis for the origin of life doesn't wash; it was merely something that had to exist for materialist reductionism to dis- place intelligent creative causation. Thus we find an inversion and variation of Voltaire's quip, "If God did not exist it would be necessary for us to invent Him." For materialists, in order for God *not* to exist, it was necessary for them to invent the soup.

That is not to say that proceeding *as if* the soup existed did not produce important scientific discoveries. Such research has given scientists a far greater appreciation of the delicacy of the conditions necessary for life as well as an immensely enhanced appreciation for biomolecular complexity. To quote Yockey again, "Although the Oparin-Haldane paradigm is now just a relic of the cosmology of the time when it was invented, it certainly de-

[39]Stephen J. Mojzsis, Ramanarayanan Krishnamurthy and Gustaf Arrhenius, "Before RNA and After: Geophysical and Geochemical Constraints on Molecular Evolution," in *The RNA World: The Nature of Modern RNA Suggests a Prebiotic RNA*, 2nd ed., ed. Raymond Gesteland et al. (Cold Spring Harbor, N.Y.: Cold Spring Harbor Laboratory Press, 1999), p. 6.
[40]See Thaxton, Bradley and Olsen, *Mystery of Life's Origin*.

served extensive research and much has been learned in investigating it. The same can be said for many other failed paradigms (Kuhn, 1957, 1970)."[41] Just as chemistry was able to gather essential chemical details under the failed paradigms of alchemy and phlogiston theory, so also attempts to reproduce the fundamental, complex prebiotic macromolecules in simulated early Earth conditions in the laboratory have yielded valuable details about the actual workings of a variety of protein structures, metabolic pathways, lipids, RNA and DNA.

It's just that the most valuable lesson of all is that such complex structures were not produced by an abiotic and purposeless shuffling of chemicals. As Yockey explains,

> It is unsupported by any other evidence and it will remain *ad hoc* until such evidence is found. Even if it existed, as described in the scenario, it nevertheless falls very far short indeed of achieving the purpose of its authors even with the aid of a *deus ex machina*. One must conclude that, contrary to the established and current wisdom a scenario describing the genesis of life on earth by chance and natural causes which can be accepted on the basis of fact and not faith has not yet been written.[42]

The problem begins as a probabilistic hurdle. If we grant the "big if," that is, if we grant the prebiotic soup rich in all the right chemical constituents, what is the probability of attaining a specific chain of one hundred amino acids by chance? As a purely mathematical exercise, we would say that, since there are twenty relevant amino acids then there are 20^{100} possible arrangements of the 100-amino-acid-long chain—about a probability of 1 in 10^{130} of getting the right sequence.[43]

And the problem only grows more daunting from there. Note that Yockey's question of probability is asked in almost total mathematical abstraction from the real biological context in which such a protein actually

---

[41]Yockey, *Information Theory*, p. 239. Yockey is here referring to Thomas Kuhns's classic *The Structure of Scientific Revolutions*. See also Hubert P. Yockey, "A Calculation of the Probability of Spontaneous Biogenesis by Information Theory," *Journal of the Theoretical Biology* 67 (1977): 396.

[42]Yockey imaginatively depicts the *deus ex machina* as one of the three Fates, the goddess Lachesis, who disposes lots in regard to chance events. Yockey represents her as churning methodically through abstract probabilities independent of factors that would render such an analysis questionable or at least greatly reduce the probability. Yockey, "A Calculation of the Probability," p. 385.

[43]We are not taking into account, in this purely mathematical exercise, the factors that would either shrink or expand the improbability, such as the harmless substitution of one amino acid for another, the problem of reaction interference, the chirality of molecules, etc.

occurs and has functional relevance. The question of probability is abstracted from time and space, as if protein synthesis could occur *ex nihilo* and *de novo*—that is, *outside* the organization of time and space provided by the living cell, and as if no factors were necessary (such as cellular protection and direction), and as if no other factors existed to dramatically lower the probability (such as chemically interfering reactions).

Actual (as opposed to abstract) protein synthesis depends on hundreds, if not thousands, of support activities and structures in the cell that are provided by a multitude of other proteins of even greater complexity than a mere 100-amino-acid chain. The cell is not simply some confined boundary where chemicals associate randomly in a kind of amorphous soup, but like Shakespeare's plays, it is tightly and elegantly organized drama in regard to time and place, where the intrinsic powers of the chemical elements are used with exquisite efficiency.

Too many scientists, driven by reductionist and materialist presuppositions, abstract a stick-figure cell from the myriad of necessary particulars in the tightly defined organization of real cells. In a real cell, a specific chain of one hundred amino acids is generated with the help of a particular and highly complex ribosome translating a particular and highly complex strand of mRNA, with the help of highly complex tRNA. Here we no longer speak of the chance generation of the amino acid sequence; rather, if nothing disrupts the translation, the specific 100-amino-acid chain will be produced—for the sake of a specific function as defined by the cell.

If such faithful translation happened only 90 percent of the time, then, given the multitude of translations continually occurring in any living cell, almost all cellular life would soon die off. The fact that life is so abundant, diverse and robust attests to the amazing fidelity of translation built into living cells. We have discovered not merely machines but robots, and not merely robots but robots capable of producing other functioning robots. And to break away further from materialist dogma, which speaks of living things as mere machines, the cells are not merely highly advanced robots but living things capable of reproducing living things.

Materialism erases the distinction between nonliving and living things, and that misses the essential nature of the way proteins exist in cells. A functional protein structure depends on the *living* unity of the cell; that is, it is both built by the cell and built for the cell, so that its structure is specified by its function as a part in relation to the living whole, one whose particular, complex arrangement of parts is necessary to carry out its intricate function.

It is important to add that this structure cannot be reduced to the coding sequences of DNA. Not only is DNA an "unread book" without the vast array
of elaborate protein structures that produce, maintain and transcribe it; but
even more, the code is only *meaningful* (i.e., functional) insofar as it produces those structures as defined by the thousands of activities orchestrated
at specific times and in specific places by the living cell. The particular complexity of a particular protein, then, includes not only its structure, but also
the specific timing and intricate place of its production and activities. The
cell runs a tight ship, and defective protein strands are quickly ground up
and recycled.

If we understand real cells, then we can place chance in its proper place,
a very subordinate place. Chance is only a remote and secondary consideration in the living cell, one that is associated with defect and destruction
rather than beneficial effect and construction. Since it is the very complex
integration of the cell that supports its activities, the "probability" of synthesizing a particular needed protein decreases dramatically as the necessary
complexity and integration *decrease*. Because the minimum threshold of
complexity and integration needed for protein synthesis is so high, only a
comparatively slight defect is needed to disrupt the cell's abilities and reduce
the "probability" of synthesizing a particular protein to zero. Important for
our purposes is that this drop to zero occurs far, far above the highest, most
hopeful heights of organized complexity climbed by origin-of-life researchers in the most artificially designed laboratory experiments.

In contrast to the reductionism of Richard Dawkins, both the order of the
amino acids and the order of the letters in Shakespeare's sentences are defined from the top down. In regard to the order of the amino acids, it is not
just that an individual ribosome is producing a particular strand of amino acids, but that the polypeptide strand is being produced for the sake of a very
specific need of the cell, one already spelled out, just as "Methinks it is like a
weasel" is already spelled out in the context of Shakespeare's entire *Hamlet*.
The context of the whole organism is essential to the existence and function
of the "text" of any amino acid. Any significant and functioning part that origin-of-life researchers have attempted to explain within a purely materialistic
framework has required other parts in order to function, thereby showing that
the essential, definitive parts of the cell exist as interdependent and cannot be
built up one blind step at a time in abstraction from the cell as functional.

It should not surprise us, then, that attempts to describe the origin of the
cell as proceeding from bottom to top, parts to whole, have indeed run into

a classic chicken-or-egg problem, the most famous of which is the DNA-protein quandary. In the words of historian and philosopher of biology Iris Fry,

> Every living cell known to us is made up of several kinds of macromolecules, including nucleic acids and proteins. Nucleic acids—DNA and RNA—store and transmit genetic information, while the proteins perform enzymatic activity, which determines all the functions of the cell. The biological synthesis and activity of nucleic acids and proteins are totally interdependent: protein synthesis is directed by the information in nucleic acids—by the specific sequences of nitrogen-containing bases in DNA and RNA; nucleic acids are synthesized, replicated, transcribed, and translated into proteins only through catalysis by enzymes. The original emergence of this "vicious circle," which clearly demonstrates the involved nature of biological organization, is a cause for wonder among biologists. Proteins and nucleic acids are extremely complex molecules, a fact that makes it hard to imagine their simultaneous synthesis on the primordial earth. And yet how could the one be produced without the other? This "chicken-and-egg" problem constitutes one of the major stumbling blocks in research on the origin of life. In fact the solution to the origin problem may be described as a resolution of the chicken-and-egg problem.[44]

Of course, the circle is only "vicious" if it is seen as poisoning the attempt to derive living things by some grand series of chemical accidents.[45] Only in the flatland of materialism is the circle inescapable, that is, vicious. Where the dimension of mind is allowed, no such prison exists.[46] In the cell there is an exact relationship of time, place and material in regard to the construction of needed proteins. In this, the cell acts not as an intelligent agent but with the same kind of parsimony and precision as an intelligent agent. An intelligent cause orders things in regard to time and place (or, if one prefers, time and space), and the structures and material used are appropriate to, and function for the sake of, the overall order. A cell does not randomly produce protein chains, day and night, out of a limitless pool of amino acids in

---

[44]Iris Fry, *The Emergence of Life on Earth: A Historical and Scientific Overview* (New Brunswick, N.J.: Rutgers University Press, 2000), p. 100.

[45]Fry's explanation is well worth quoting precisely because she is not a theist-friendly critic of origins research but remains faithful to the reductionist project.

[46]For an overview of the state of origins research written from a nontheistic viewpoint, see Fry, *Emergence of Life;* Lahav, *Biogenesis;* and Yockey, *Information Theory.* Two surveys that argue for intelligent design are Thaxton, Bradley and Olsen, *Mystery of Life's Origin;* and Rana and Ross, *Origins of Life.* See also Shapiro's excellent *Origins;* Klaus Dose, "The Origin of Life: More Questions than Answers," *Interdisciplinary Science Reviews* 13, no. 4 (1988): 348-56; and Leslie Orgel, "The Origin of Life: A Review of Facts and Speculations," *Trends in Biochemical Sciences* 23 (December 1998): 491-95.

hopes that it will happen on the one it needs, any more than Shakespeare
had by his desk a bulging sack of Scrabble letters into which he continually
dipped in hopes of randomly producing meaningful drama, letter by tortuous letter.

The hope that such random production could provide the required complex organization of life simply isn't based on evidence. In the abstract, we
might think that it is possible we could fill not just our ponds, but all our
oceans with amino acids; let them swirl on the early Earth; and surely, given
the size of the oceans and a billion years, we could get that 100-amino-acid
sequence *somewhere* and *sometime*. But as Sean Taylor and his colleagues
have calculated in regard to a protein one hundred amino acids long,[47]
"Even a library [of amino acid chains] with the mass of the Earth itself—5.98
x 10^{27}g—would comprise at most 3.3 x 10^{47} different sequences," representing only a "miniscule fraction" of the full number of possible amino acid
chains of 20^{100}. The same hurdle faces other highly complex and necessary
cellular parts. We need an impossibly large bowl of prebiotic soup, and we
can't even find evidence on the early Earth for a relatively small one.

Researchers often get around this problem by what amounts to a kind of
intellectual sleight of hand. Robert Shapiro explains:

> Prebiotic chemists . . . assume that a substance, once demonstrated in any
> amount as a product of a prebiotic reaction [in the laboratory], can then be
> employed in pure form and enhanced quantity as the starting material in a
> very different prebiotic transformation [again, in the laboratory]. This process
> is repeated until an entire series of reactions has been put together, to connect
> [for example] the reducing atmosphere with a replicator.[48]

The researchers, who obviously realize that chance needs an overflowing
cornucopia of material, supply it in "pure form and enhanced quantity" as
the starting point of each attempt to demonstrate that chance can climb the
next level of complexity—that is, in an amount that their original experiment
most decidedly did *not* produce and in such purity and concentration as
could never be arrived at or sustained by natural means (and would be quite
evident geochemically if it had).

A related error—again, aptly noted by Shapiro—is that, in the step-by-
chemical-step generation of large biomolecules, such researchers link to-

---

[47]Sean V. Taylor et al., "Searching Sequence Space for Protein Catalysts," *Proceedings of the National Academy of Sciences*, 98, no. 19 (2001): 10596.
[48]Shapiro, *Origins,* p. 177.

gether levels of complexity as if they had occurred consecutively on the early Earth even though each experiment, representing each step up in chemical complexity, was run under very different conditions. To use again the attempt to produce a replicator, a researcher will begin with a Miller-Urey type experiment, producing hydrogen cyanide and formaldehyde. When formaldehyde is exposed to another set of conditions, it yields a mixture that includes the sugar ribose; when the hydrogen cyanide is exposed to very different conditions, it produces some adenine.

Shapiro continues, noting that other laboratory procedures, much more indirect and complicated, can produce the other bases of the nucleic acids. Next, adenine and ribose, under very different conditions, can be heated—with the proper catalysts—to get adenosine, thereby taking care of a component of RNA. Further, adenosine can be heated with phosphate and another array of appropriate catalysts to give us a nucleotide, and if other procedures are used, those nucleotides can be connected together and we've as good as got a replicator.[49] Voilà!

The problem, Shapiro explains, is not only that the yield from one step is grossly insufficient to supply the material for the next level of chemical transformation, but that since the transformations occur under very different conditions, the conditions that allow for one step either do not allow or actually inhibit the next.[50] Nature could never be as nimble as the intelligent agents who designed these lab-synthesized replicators.

But just suppose (to return to the example of protein), that a particular protein which normally exists in a cell was generated *de novo*. Would it be functional? Let us use an example to show how it might appear to be functional in the abstract but, on closer inspection, is not. In a famous experiment by C. B. Anfinsen, it was found that an unfolded (or denatured) protein, the enzyme ribonuclease, would refold itself correctly, gaining its original three-dimensional form and functional catalytic activity in under a minute.[51] Obviously, as Franklin Harold notes, "The amino acid sequence is

---

[49]Ibid., pp. 176-78. We've shortened Shapiro's delightful account, as put into the mouth of Dr. Midas, who "can convert mundane chemicals into genes with a wave of his hand and a well-chosen phrase."

[50]If one tries to circumvent this problem by stretching the steps over time so as to allow for a change of conditions, another set of problems arise. How, for instance, would a sufficient supply of (say) the sugar ribose remain intact long enough to function as a backbone for nucleic acids under the conditions that also bring about its transformation into adenosine?

[51]A short description can be found in Harold, *Way of the Cell*, p. 50. See C. B. Anfinsen, "Principles that Govern the Folding of Protein Chains," *Science* 181, no. 96 (1973): 223.

fully sufficient [in this case] to determine that protein's three dimensional configuration, and also its biological activity."[52]

Or, to take another example, Robert Allen discovered that cellular vesicles (which are membranes acting like little cargo boxes used to transport substances from one place to another within the cell) continued to move on a determinant path along microtubules even when removed from the living cell.[53]

Magical! Yes, pixie dust materialism all over again, akin to the irrational belief that cells were just little blobs of protoplasm.[54] The same goes for these particular "parts" themselves. A few years after Allen's discovery, researchers learned why cellular vesicles were so strangely cooperative in their movement: the vesicles were being shuttled along microtubule tracks by tiny molecular motors (a protein molecule named kinesin).[55] If these molecular motors were produced by chance, and if by some other prodigious miracle of chance a microtubule was also randomly produced right next to it, they would still be as useless as an empty coal train on a track to nowhere. The function of the dense matrix of microtubular tracks in a cell is to allow these particular molecular motors to deliver particular chemical substances in the cell from one place to another. Outside of that context, a kinesin's activity is useless.

As we explore the cell, such instances multiply. To return to Anfinsen's experiment, the significance of protein folding is both overrated and misunderstood. Biologists too easily assume, based on experiments like Anfinsen's, that "interactions occurring within and between polypeptides are all that is necessary for the biogenesis of proteins in their functional form." However, as Wayne Becker, Lewis Kleinsmith and Jeff Hardin point out, "This model for self-assembly in vivo (in the cell) is based entirely on studies with isolated proteins, and even under laboratory conditions, not all proteins regain their native structure [after denaturation]."[56]

---

[52]Harold, Way of the Cell, p. 50.

[53]Robert Allen et al., "Fast Axonal-Transport in Squid Giant-Axon," 218, no. 4577 (1982): 1127-29.

[54]The belief that the microbiological realm was relatively simple was held against strong evidence, even before powerful microscopes, because a straightforward application of reverse-engineering thought (a consideration of what biologists already knew the microbiological realm accomplished) pointed to a microscopic realm of great intricacy.

[55]Boyce Rensberger, Life Itself: Exploring the Realm of the Living Cell (New York: Oxford University Press, 1996), pp. 31-37.

[56]Wayne Becker, Lewis Kleinsmith and Jeff Hardin, The World of the Cell, 4th ed. (San Francisco: Benjamin Cummings, 2000), p. 32.

And while self-assembly may be sufficient for simpler protein structures, the more complex the structure, the more cellular guidance is needed, in great part because the amino acid sequence alone could result in several different folding patterns, only one of which is functional. One interesting form of help from the cell occurs through another type of protein, the chaperone. In Boyce Rensberger's words, chaperone proteins "grasp the new, partially folded chain and push or pull it into one particular shape. Like a sculptor modifying an unsatisfactory clay figure, the chaperones massage the amino acid chain, nudging a helix sideways or shifting a loop from this side to that."[57]

Protein folding is therefore misunderstood as some kind of vindication of the powers of chemicals to create living complexity on their own. Imagine the following scenario. You enter Professor Mirandus's laboratory with a reductionist colleague. Mirandus welcomes both of you warmly, and announces, "You've come just in time, friends. Watch this!" He carefully steps over a line of several hundred magnetic boxes (about the size of small matchboxes, but of many different shapes) strung together with wire and lying on the floor; he throws a switch; and suddenly the line of magnetic boxes begins to fold in a very orderly way into a dense but elegant tangle. Then, this newly created tangle immediately begins to clean up the laboratory. You watch in speechless astonishment, and when the last polished beaker is put carefully away, Mirandus turns the electricity off, and the boxes fall into a lifeless pile on the floor. Still stunned, barely able to speak, you look over at your colleague. You're amazed to see that he's wearing a studied look of indifference. "Don't you think that's amazing?" you blurt out. "Not really," he answers, "it's just a bunch of magnets."

Now many of the activities carried on by protein structures in the cell are far more amazing than Professor Mirandus's laboratory-cleaning magnet-enzyme: consider the RNA polymerase that transcribes specific, needed information from the DNA to make a strand of mRNA; the signal proteins that somehow find the proper gene on the enormously long and complex DNA strand so that RNA polymerase will indeed transcribe what the cell has signaled a need for; or the repair enzyme that works its way along a newly made strand of DNA checking for errors and "ripping" out segments that contain errors so that DNA can make a correct replacement. Can we say of these, and a hundred other specific protein structures carrying out equally intricate tasks, "They're just a bunch of amino acids"?

---

[57]Rensberger, *Life Itself,* p. 105.

The proper, natural response is astonishment that the structure and chemical potentialities of hydrogen, nitrogen, carbon and oxygen, as taken up into the various amino acids, can be formed into a structure with the power to perform complex and highly specified activity—a structure integrated into, and for the sake of, an even larger whole, the living cell. Ironically, no one, not even the crassest reductionist, would glibly say that "it's just a bunch of magnets" after witnessing Professor Mirandus's laboratory-cleaning magnetenzyme, yet the multitude of activities carried on by proteins in the cell are far more elaborate and precise. Moreover, and even more telling against reductionism, the proteins fulfill their essential tasks because they are tightly controlled from the top—integrated into the hundreds, even thousands, of chemical reactions and activities that define the daily life of even the most ordinary cell.

We need to remain clear about just how large the proper perspective is. The very birth of the chemical elements hydrogen, nitrogen, oxygen and carbon depended on fine-tuning. Now we find that these same elements appear exquisitely designed for a structural self-assembly machine, resulting in what is akin to an intricate army of industrial robots, perfectly fit for the multitude of tasks in the great factory of the living cell. We should note, moreover, that in comparison to our clumsy factory robots, the living robots of the cell represent exponentially more complex activity at a size exponentially smaller than our most advanced nanotechnology.

All this, of course, makes all the more ludicrous the claim that proteins or any cellular structure could have arisen by a purely undirected means. This becomes clearer still when we realize just how many distinct cellular parts exist in even the simplest cells. If we look at the humble but famous bacterium *Escherichia coli* (also known as *E. coli*), we find this single-celled creature to be anything but simple. As Harold writes, "What the eye beholds depends on what it looks through." Looking through the comparatively primitive microscopes of Darwin's day, the cells do appear as nearly formless, simple blobs. Using the latest instruments and techniques of our own day, biochemistry has revealed a miniature cosmos. "There are more than 2 million protein molecules per cell, potentially of four thousand kinds, and nearly a thousand species of small molecules; 300 million molecules in all, not counting water which makes up nine tenths of the cell's mass," Harold explains. "All these jostle one another in a [cellular] volume of about one cubic micrometer."[58]

---

[58]Harold, *Way of the Cell*, p. 66.

If we hoped to find a cell "simpler" than *E. coli*, the prospects for its chance generation still remain bleak. We can't get much simpler. As philosopher of science Stephen Meyer notes, calculations of the minimal complexity needed for cellular life put the lower limit somewhere between 250 and 400 genes and their corresponding proteins.[59] Such is the minimal number of parts necessary for functioning and, hence, the minimum that allows *any* part to exist as a functioning, specified part.

And so the living cell appears, again, happily unscathed by the reductionist attempts to make it disappear into a few of its details. It remains a living whole, unable to be built up one blind step at a time because, like the dramas of Shakespeare, it is not the mere sum of its parts. Nature has spoken most eloquently. The failure of origin-of-life research (or rather its unexpected finding) means the failure of the entire materialist, reductionist paradigm. For if a single cell cannot be built up via materialist assumptions, then the entire edifice of life depends on something more.

The failure of reductionism, however, isn't restricted to the bottom of nature. As will become clear in the next chapter, reductionism fails all the way up the line, from the microscopic realm, to the world of animals and humankind, to Shakespeare and beyond.

---

[59]Stephen Meyer, "DNA and the Origin of Life: Information, Specification and Explanation," in *Darwinism, Design and Public Education,* ed. John Angus Campbell and Stephen C. Meyer (East Lansing: Michigan State University Press, 2004), p. 243.

9

THE RESTORATION OF THE
LIVING ORGANISM

And all shall be well and

All manner of thing shall be well

When the tongues of flame are in-folded

Into the crowned knot of fire

And the fire and the rose are one.

T. S. Eliot, "Little Gidding"

RETURN NOW TO PHYSICIST STEVEN WEINBERG'S COMMENT that
the more we understand scientifically about the universe, the "more it also
seems pointless," with humankind "just a more-or-less farcical outcome of a
chain of accidents reaching back to the first three minutes." Again, his claim
is that, as a matter of scientific fact, we indeed have demonstrated that the
universe is meaningless. Ironically, Weinberg also believes that science offers a kind of saving grace, salvaging meaning from meaninglessness by
building "telescopes and satellites and accelerators" and sitting at "desks for
endless hours working out the meaning of the data they gather." As he says,
"The effort to understand the universe is one of the very few things that lifts
human life a little above the level of farce, and gives it some of the grace of
tragedy."[1] How odd that there could be such a thing as "the meaning of the
data" if the universe itself is a pointless chain of accidents.

In the last chapter we saw a similar irony played out in regard to the origin of life. Most origin-of-life researchers took it for granted that the uni-

---

[1]Steven Weinberg, *The First Three Minutes: A Modern View of the Origin of the Universe* (New
York: Basic, 1977), p. 154.

verse in all its detail is the product of blind forces. On this view, since there is no ultimate intelligent cause, the order created is accidental, and hence no ultimate meaning can be derived from it. All seemingly real and ordered wholes—stars, planets, trees, cats, finches, cows and humans—are illusory, reducible to a chain of accidents. The only goal of science is to take the wholes apart—piece by piece, smaller piece by smaller piece, until it has read the smallest parts—and then to deduce by what aimless road the smallest particles came to produce the absurd results (absurd, not *tragic,* since tragic results aren't possible in a world drained of meaning). On this absurd view, nature is readable (oddly enough), but the cosmic tale is, to paraphrase Macbeth, told to us by an idiot, full of sound, fury and energy, and in the end (as well as the beginning) signifying nothing.[2]

Macbeth was tragic when he spoke such words. But he had so violated the moral order, so given himself over to forces of darkness, that he imagined that all was an amoral chaos. Of course, William Shakespeare did not view Macbeth's situation as Macbeth did; otherwise he would not have written a tragedy about him. In Shakespeare's universe, justice is not always swift, but it is sure, and from this conviction he fashions a dramatic world that grinds the villains "exceedingly fine." Unlike Macbeth under Shakespeare, humankind under materialism is but a purposeless shuffle of atoms in the void. There can be no human tragedy without the background of cosmic purpose for the same reason that the highest development of dramatic character cannot be achieved without plot.

Hamlet is tragic, but the same characters in Tom Stoppard's *Rozencrantz and Guildenstern Are Dead* are merely silly and clueless, and that's the work's point: Stoppard understands that humans under materialism are not tragic; they are only absurd. The absurd characters in Samuel Beckett's *Waiting for Godot* are waiting for God. He never comes because, on the materialist view, he never was. When sitting through a showing of a well-wrought piece of absurdist drama, the theist perceives a tragedy; but it's the tragedy of a culture falling into darkness. The post-theist feels it too, one would guess, because the God of the West has not yet wholly been flushed from her system.

The dramatic experience provided by such works, then, can be powerful stuff. Absurdist theater may not touch the deepest currents of being, but it does describe with arresting clarity a nihilistic subculture given over to the tenets of materialist reductionism.

---

[2]Shakespeare *Macbeth* 5.5.30-31.

However, as we saw in the previous chapter, when this philosophical dogma finally bumped up against the hard evidence of the microscopic world, it failed. Living cells cannot be built up from a chance aggregation of parts. They must be put together by a means much more analogous to Shakespeare's creation of *Hamlet*.

Again, this isn't to say that materialist reductionism hasn't borne fruit. As the history of chemistry amply attests, the history of scientific inquiry offers a multitude of examples of useful but ultimately false scientific theories, theories that helped advance science but which were ultimately discarded because the order of nature did not conform to the dictates of the theory. We now know that the philosopher's stone and dephlogisticated air exist only in the imagination: there is nothing in reality that corresponds to these terms. They have meaning only for historians of chemistry. The same is true, we argue, for the terms *prebiotic soup, RNA world, chance formation of functional proteins* and a number of other theoretical entities and phrases postulated by the reductionist hypothesis regarding the origin of life. They directed research for a time and in so doing showed that there is nothing in the physical world that corresponds to them. They are the hopeful monsters of the hopeful materialist imagination.

Thus it is that the failure of materialism is not merely negative, for its failure highlights meaning in the natural world. The failure to reduce the cell to its material parts, for instance, reestablishes the primacy and reality of the living cell and, indeed, reestablishes biology as the study of living things. But even more, the failure of materialist reductionism as a mode of science reestablishes the study of living things as they present themselves, including that most curious of creatures, human beings. In so doing, it restores the study of human geniuses like Shakespeare, who so aptly and deeply probed and represented the human drama in its dazzling complexity. Ironically, the restoration of genius (as a concept) rests on the restoration of our common sense.

That the cell is alive is, of course, a commonsense notion, based on our experience of much larger, multicellular animals. As large animals ourselves, we have historically reasoned from cows to cells, from the visible to the microscopic. That such reasoning has a firm basis in reality is being affirmed not just negatively by the failure of origin-of-life reductionism, but positively by the new structuralist approach in biology, a top-down, whole-to-parts way of studying living things.

According to structuralism, the parts exist not merely as historical aftereffects of the unpurposed association of yet-smaller material parts, but as recognizable, functional entities understood in light of the living whole to which they belong. One of the reasons for the structuralist rejection of reductionism is that, contra Charles Darwin, there are great leaps in nature, beginning with the great leap between nonliving and living things. Franklin Harold writes,

> Even the simplest unicellular creatures display levels of regularity and complexity that exceed by orders of magnitude anything found in the mineral realm. A bacterial cell consists of more than three hundred million molecules (not counting water), several thousand different kinds of molecules, and requires some 2,000 genes for its specification.

He is on the verge of a heresy against Darwinism, but he drives on nonetheless:

> There is *nothing random about this assemblage,* which reproduces itself with constant composition and form generation after generation. A cell constitutes a *unitary whole,* a unit of life, in another deeper sense: like the legs and leaves of higher organisms, its *molecular constituents have functions.* Whether they function individually, as most enzymes do, or as components of a larger subassembly such as a ribosome, *molecules are parts of an integrated system,* and in that capacity can be said to *serve the activities of the cell as a whole.* As with any hierarchical system, each constituent is at once an entity in itself and a *part of the larger design;* to appreciate its nature one must examine it from both perspectives. Organization, John von Neumann once said, has *purpose;* order does not. Living things clearly have at least one purpose, to perpetuate their own kind. Therefore, organization is the word that sums up the essence of biological order.[3]

The essence of biological order is not the nonliving part, then, but the living whole into which the part fits and makes sense. The chemical elements themselves (made possible by fundamental, cosmological fine-tuning) point toward living things, carrying in their very structures extraordinary and exact chemical potentialities beautifully designed for actualization in the biological world. These chemical potentialities of elements and compounds only become actual and knowable when they are taken up into a more complex form. As we have said, no one could know, just given hydrogen by itself, the crucial features it brings to H_2O (the liquid of life) or to the intricate folding of proteins; nor could anyone guess the amazing properties of

---

[3]Franklin Harold, *The Way of the Cell: Molecules, Organisms and the Order of Life* (Oxford: Oxford University Press, 2001), pp. 10-11 (emphasis added).

water given just water, for they are revealed only in the living context of water as the liquid of life.

But nature appears fitted for biology in more than just its physical constants. The form of the cell tightly integrates its chemistry to serve its ongoing life (including its reproduction). As should be clear from the last chapter,
this kind of integrated, purposeful, top-down ordering of space, time and
material functions in exactly the opposite way to the purposeless ordering
of chance. Indeed, the molecular parts of a cell work in concert like an elaborate, operatic staging of Shakespeare. Boyce Rensberger calls a cell's nucleus "a center of a veritable symphony of chemical signals arriving from all
directions." Its behavior depends on these outside influences. And the signaling goes both ways. "The life of a multicelled organism depends on an
extraordinarily complex interplay of thousands of different chemical signals
coming from each cell and going to some or all of the other cells that make
up the republic of cells that is the human body." He continues:

> Whatever the source of the signal, its effect is to work within the nucleus,
> seeking out the regulatory portions of DNA that govern the specific gene it
> "wants" to activate. The signal proteins must find and bind to the regulatory
> segments to cause a given gene's message to be "expressed" or, depending on
> the signal, to prevent expression. . . . At one time it was assumed that the sig
> nal proteins wandered aimlessly within the nucleus until they chanced to
> bump into their target sequence. Now it is clear that they don't. They grab onto
> a DNA strand and "walk" along it, "looking" for the sequence—yet another ex
> ample of the role of autonomous motion in life. Even though the molecule
> may have to walk a long way to find its gene . . . the process is roughly 100
> million times more efficient than simply bouncing around inside the nucleus.[4]

Note that, in accord with the dictates of materialist reductionism, not only
did many biologists assume that the first organism could have arisen by aimless chemical association, but they also assumed that "the signal proteins
wandered aimlessly within the nucleus until they chanced to bump into their
target sequence."

An obvious sign of the irreducible reality of the living whole is that the
cell not only governs its own complex internal chemical activity for the sake
of its ongoing life and reproduction, but that it acts as a "unit of life." Michael
Behe has argued forcefully that the bacterial flagellar motor needs all of its

---

[4]Boyce Rensberger, *Life Itself: Exploring the Realm of the Living Cell* (New York: Oxford University Press, 1996), pp. 93-94.

parts to function,[5] but even more amazing is the way a bacterium uses this part for the sake of its own needs. The justly famous *E. coli* uses its flagellum in an interesting pattern of swimming in a straight line and "tumbling" to change direction so as to find food in whatever medium it happens to find itself. Experiments have revealed that

> *E. coli* can respond within milliseconds to local changes in concentration [of the nutritional attractant], and under optimal conditions readily detects a gradient [in the nutritional attractant] as shallow [i.e., as diluted] as one part in a thousand over the length of the cell. Cells remain responsive to attractants over a concentration range of five orders of magnitude, nanomolar to millimolar. In effect, each cell performs a continuous series of rapid computations and acts upon their output.[6]

Such "computations" are done by the living bacterium for the sake of finding food, and it uses its flagellum *for the sake of* this end. The flagellum doesn't just make swimming and tumbling motions pointlessly.

The same kind of primacy of the whole exists in multicellular organisms, from cats to cows to humans. The cells of a multicellular organism are alive—just as a single-celled bacterium is alive—and many can be taken out of the body and exist as fascinating single-celled creatures, crawling around on the bottom of a cultured petri dish (albeit provided with nutrients, oxygen, growth factors, etc., which the complex organism previously supplied) and dividing as well. However, the form of the whole—the creature from which they were derived—seems even in this state to be still imprinted on them, so that under the proper conditions they take up their specialized cellular task again (skin cells, e.g., forming a sheet on the bottom of the dish, or fibroblast cells continuing to produce collagen, or female breast cells manufacturing and secreting milk protein.)[7] But the really strange thing—strange, though it is central to complex life—

---

[5]Michael Behe has argued that the bacterial flagellum has the signature of an assemblage of parts purposefully arranged; see *Darwin's Black Box* (New York: Free Press, 1996). He answers objections in "Self-Organization and Irreducibly Complex Systems: A Reply to Shanks and Joplin," *Philosophy of Science* 67 (March 2000), available online at <http://www.discovery.org/scripts/viewDB/index.php?command=view&id=465>. Scott Minnich and Stephen Meyer describe experimental evidence further corroborating Behe in "Genetic Analysis of Coordinate Flagellar and Type III Regulatory Circuit in Pathogenic Bacteria," in *Design and Nature II: Comparing Design in Nature with Science and Engineering*, ed. M. W. Collins and C. A. Brebbia (Boston: WIT, 2004), available online at <http://www.discovery.org/scripts/viewDB/index.php?command=view&id=2181>.

[6]Harold, *Way of the Cell*, p. 89.

[7]Rensberger, *Life Itself*, pp. 15-17.

is that an animal's tens of trillions of cells (comprising some 200 different kinds) act in concert according to the whole animal over distances in the organism many, many magnitudes greater than the size of the cells. The same kind of unifying teleology that exists in a single cell exists among cells in the multicellular organism.

The teleology is seen vividly in the activity of growth, where the function of cells and the organs of which they are constructed are determined long before they can actually function in the organism—as if the living being-to-be were exerting an organizing force, according to its ultimate structure or form, on the cells in the developing organism. To use humans as an example, during the first month of embryonic formation, before a human form is recognizable, cells have divided and organized to form the neural tube, the inchoate embryonic spinal cord and brain. As Rensberger explains, a closer look at the process reveals the purposeful orchestration of cells for the sake of forming the entire organism:

> As the [neural] tube closes, one of the most extraordinary classes of cells develops along the crests on each side of the neural groove. Just before the crests meet to close the tube, these so-called neural crest cells jump off and crawl away to pursue their own fates in other parts of the embryo. . . .
>
> These cells migrate off to an unusually wide variety of tasks, some to become various parts of the nervous system (including all the sensory nerves), others to help construct the teeth, still others to make some forms of cartilage and the bones of the face, and yet another group to form parts of the adrenal glands. Some neural crest cells migrate out to what will become the skin, where they take up permanent residence and manufacture the melanin that gives skin and hair its color. The broad array of very different roles in this drama has long astounded and perplexed developmental biologists, for when the cells leave the neural tube, they all appear to be identical, unspecialized cells. Yet after they take up their respective duties throughout the organism, each type of specialized cell has its own characteristic suite of active genes. The genes for making melanin, for example, are shut off in neural crest cells that give rise to bone; and the bone-making genes are disabled in melanocytes, the melanin-making cells.[8]

We must remind ourselves that all of this orchestrated diversity springs from a single fertilized egg, containing a single "text" of DNA. The DNA is not the ultimate cause of biological formation; rather DNA is the informa-

---

[8]Ibid., pp. 179-80.

tional material used by the growing organism (just as words are the material for Shakespearean drama). More and more, biologists are concluding that the living whole is the true organizing reality, not the DNA, a mere part. Paul Nelson and Jonathan Wells are particularly lucid on this point. In "Homology in Biology," they discuss how structuralists are downgrading the role of DNA precisely because it isn't the magic materialist genie that causes everything to happen. To illustrate the organismal context principle, they offer the analogy of a lexicon:

> Any text can be broken down into a lexicon comprising its words. For instance, the concluding lines of the Gettysburg Address constitute a 43-word lexicon. Using only that lexicon, one can then write a completely different text—for instance, an anarchist's manifesto. Here the meanings of single words and phrases, namely, their functional roles, are determined "top-down," by the higher level context of the paragraph; that is by the author's purpose.[9]

The sample manifesto that Nelson and Wells then create reads,

> by this we highly resolve that we shall have freedom from this nation—that devotion shall perish. These people honored the last government in vain. The dead increased. Measure thy full devotion! The earth under here gave these people birth, not a dead God, and from that they shall take their new cause, for which people have not died.[10]

Those are the very same words used for the closing lines of the Gettysburg Address, but organized according to an entirely different purpose!

In the organism, the role of order and context is even greater than in this example. Consider neural crest cells. They all contain the same genetic lexicon, but this information becomes functionally active or disabled according to their destination. Using neural crest cells from a chick embryo, researcher Nicole Le Douarin found that when removed from the chick and grown in culture, some of these cells were committed quite early to their specialized function, while others were not but could take on a variety of functions. As other research discovered, those cells that remained unspecialized became specialized as they crawled over the embryo to their destination, the unneeded genes disabling as they approached their goal.[11]

---

[9]Paul Nelson and Jonathan Wells, "Homology in Biology," in *Darwinism, Design and Public Education*, ed. John Angus Campbell and Stephen C. Meyer (East Lansing: Michigan State University Press, 2003), pp. 310, 317.
[10]Ibid.
[11]Rensberger, *Life Itself*, p. 180.

To return to the development of the human embryo, by the end of the eighth week, when the embryo is about an inch and a quarter long, "the foundation of all organs and structures have been laid down."[12] But in many cases, these foundations are laid down long before the organ itself will ever be functional, so that the development is for the sake of the whole living being. For example, before birth, the baby receives oxygen and nutrients through the umbilical cord, but if she is to survive outside the womb, she must receive them from her own lungs and liver. The lungs and liver, having developed much earlier, must now undertake their proper function; this involves an intricate, last-minute reconstruction of the circulatory system, beginning with the constriction of the umbilical arteries and then the inflation of the lungs (where the oxygen causes lung cells to release a protein, bradykinin, that in turn causes blood vessels that previously allowed blood to bypass the liver to shut off, so that blood will flow to the liver).[13]

Note the insuperable problem this poses for the materialist belief that such complex development evolved step by step, without any goal "in mind," as a set of increasingly complex but randomly generated sequence of DNA codes (augmented now and again by lateral gene transfer from other organisms, themselves struggling to evolve step by step). Once the embryo of some hypothetically evolving early animal form is ready for its life outside the womb, with all of this exquisitely orchestrated series of development tasks behind it, it is not free only then to begin evolving the capacity to switch over the circulatory system. Without this last step, all of the other steps are moot: the newborn couldn't possibly survive. Thus we see that embryonic development is not a cobbled together patchwork of evolutionary history; it is a well-directed synthesis of parts for the sake of the developed living structure or form.[14] Here again we encounter irreducible complexity—now at the higher level of the embryo, the level of living

---

[12]Ibid., p. 182.

[13]Ibid., p. 187.

[14]The thirteenth-century theologian and philosopher Thomas Aquinas (c. 1224-1274) captured the strangely goal-directed quality of the natural world long before it had been affirmed by our modern understanding of embryological development: "Hence it is clear that nature is nothing but a certain kind of art, i.e., the divine art, impressed upon things, by which these things are moved to a determinate end. *It is as if the shipbuilder were able to give to timbers that by which they would move themselves to take the form of a ship*" (*Commentary on Aristotle's Physics,* trans. Richard J. Blackwell, Richard J. Spath and W. Edmund Thirlkel [London: Routledge & Kegan Paul, 1963], lect. 14, n. 268, emphasis added).

biological form. To recall Shakespeare, the drama of growth is determined by the overall plot.

In refusing to see the whole as anything more than the sum and interaction of the parts, reductionism has therefore neglected "the problem of biological form."[15] Gerry Webster and Brian Goodwin, as part of the growing structuralist school in biology, are questioning not just some minor aspect of current neo-Darwinist thinking, but "the adequacy of the evolutionary paradigm in relation to its failure to provide any satisfactory theory of the production and re-production of biological form."[16] The failure of such reductionism is leading Webster, Goodwin and other structuralists to a new appreciation of the biological thinkers, prior to and contemporary with Darwin, whom Darwin is generally understood to have overthrown, such as Baron Georges Cuvier (1769-1832), Richard Owen (1804-1892) and Étienne Geoffroy St. Hilaire (1772-1844), as well as "a number of the 'fringe' figures outside the 'mainstream' of twentieth century biology including [William] Bateson, D'Arcy Thompson and especially [Hans] Driesch and [Conrad] Waddington."[17]

The new school of structuralism does not, of course, reject study of the material parts of organisms; rather, it understands "the problem of biological organization, and therefore of form," to be "the primary problem and questions of material composition" are therefore "secondary."[18] This focus on structure as primary has emerged not out of theistic discomfort with materialism, but from growing evidence-based dissatisfaction with the reductionist focus in orthodox neo-Darwinian biology, especially the focus on DNA as a fully sufficient cause of living beings.[19]

The belief that genes are sufficient explanations for biological order is not the result of thorough research, but rather of the reductionist paradigm itself. More recent research (not tethered to the paradigm) into what genes actually do—rather than what the paradigm dreams they must do—makes it more and more clear that while genes are necessary to biological order, they are

---

[15]G. Webster and B. C. Goodwin, "The Origin of Species: A Structuralist Approach," *Journal of Social and Biological Structures* 5, no. 1 (1982): 15.

[16]Ibid., p. 16.

[17]Ibid., pp. 16-17. See also A. J. Hughes and D. M. Lambert, "Functionalism, Structuralism and 'Ways of Seeing'" *Journal of Theoretical Biology* 111 (1984): 798.

[18]Webster and Goodwin, "Origin of Species," p. 17.

[19]For an introduction to the variety of antireductionist researches going on in biology today, see the collection of essays in *Beyond Neo-Darwinism: An Introduction to the New Evolutionary Paradigm*, ed. Mae-Wan Ho and Peter Saunders (London: Academic Press, 1984).

far from being sufficient.[20] In the words of biologist H. F. Nijhout, "The simplest and also the only strictly correct view of the function of genes is that they supply cells, and ultimately organisms, with chemical materials. These materials can be the gene products themselves, but often they are things made, altered, or imported by the gene products." Far from being the great hoped-for genie of materialism, "genes are passive sources of materials upon which a cell can draw."[21]

Some might think that this description lacks the romantic sweep present in the PBS-style theater that equates the discovery of the double helix with the "secret of life." The structuralist account has only plain accuracy to recommend it: DNA isn't the secret of life. The problem with narrowly focusing on genes is that, while they supply needed material, they do not specify, in accordance with place and time, what material is needed and when and where it is needed. These specifications occur only in and because of the integrated structure of the living cell—the very structure or form that reductionism takes to be specified by genes. Thus Goodwin argues, "Genes act within an organized context [i.e., the cell and its substructures] whose dynamic order must be exactly defined (and not explained away as a result of 'complex interactions' or 'pleiotropy') before we have a science of morphogenesis," that is, a science of the development of biological form.[22] In short, biology must recover the rather everyday notion of the primacy of the organism. Or as Goodwin puts it in a bit more technical way, biologists must "recover the notion of an integrated system whose space-time dynamic involves the whole cell, and extends to the whole developing organism."[23]

Putting the emphasis back on the whole organism not only returns biology to the study of living things in all their wondrous complexity, but also brings sanity back to biological classification, or taxonomy. For Darwin, the focus was on individual variations and, hence, on what was peculiar or particular. Biological form was therefore considered fluid and merely accidental, render-

---

[20]For an amusing (if depressing) account of the all-pervasive power attributed to genes as a kind of cultural-scientific phenomenon, see Jon Beckwith, "The Hegemony of the Gene: Reductionism in Molecular Biology" in *The Philosophy and History of Molecular Biology: New Perspectives,* ed. Sahotra Sarkar (Dordrecht: Kluwer Academic Publishers, 1996), pp. 171-83. See also Lenny Moss, *What Genes Can't Do* (Cambridge, Mass.: MIT Press, 2003).

[21]H. F. Nijhout, "Metaphors and the Role of Genes in Development," *BioEssays* 12, no. 9 (1990): 444.

[22]Brian Goodwin, "What Are the Causes of Morphogenesis?" *BioEssays* 3, no. 1 (1985): 33. Pleiotropy is a rather fancy name that simply means that one gene can have multiple effects on the organism, or "phenotype."

[23]Ibid., p. 35.

ing classification in terms of form a questionable process. But as Hans Driesch pointed out at the beginning of the twentieth century, the Darwinian belief that organisms were produced by chance, "at once did away with any deeper meaning for zoological classification. . . . The totality of living forms appeared . . . as *meaningless* as, say, the forms of clouds in their accidental peculiarity."[24] *That* should sound familiar, for it is just such untethered reductionism that leads contemporary Darwinist Richard Dawkins to compare species to cloud formations in his famous "Me thinks it is like a weasel" discussion.[25]

He received this view directly from his intellectual progenitor. For Darwin, the focus is always on the particularity, the peculiarity, the individual variation; the species itself is not real but is, rather, an unintended and temporary appearance on a continuum of continual, particular variations.[26] This is not a mere caricature foisted onto the old Victorian by contemporary structuralists. In Darwin's own words,

---

[24]Webster and Goodwin, "Origin of Species," p. 26 (emphasis added).

[25]While structuralists have rightly pointed out the distortions caused by focusing only on the material parts of organisms at the expense of form, their occasional forays into origins science often involve a kind of reductionism in regard to form. Structuralists like Goodwin have suggested that the laws of physics and chemistry may provide the self-organizing context of complex biological form, arguing that such laws provide general parameters or "fields of force" that guide biological development along a set number of organizational pathways. While structuralists are able to provide many concrete, scientific criticisms showing the failure of neo-Darwinian reductionism, they provide scant evidence that regularities of chemistry and physics could somehow provide the larger parameters that lead to the development of complex, living biological forms. See, for example, Brian Goodwin, *How the Leopard Changed Its Spots: The Evolution of Complexity* (New York: Charles Scribner's Sons, 1994), which relies on chaos theory, mathematical modeling, and on only a few examples from actual biology, such as the development of large, singe-celled plants, leaf-budding patterns and limb bud morphogenesis—a far cry from, say, an explanation of the development of the circulatory or nervous system, or embryonic development, or the complex integration of the immune system. One problem with such explanations is that the form cannot be the result of larger physical and chemical regularities because the regularities to which structuralists appeal (often rather vaguely) have their effect only because of the complex spatial and temporal ordering provided by the cell or multi-celled organism. We should add that in e-mail correspondence (on Nov. 19, 2005) Goodwin noted that his views have continued to develop and that he agrees at least with our focus on meaning, a concept he emphasizes in his book currently in progress, *The Honeyguide to the Hive of Being,* which includes a chapter on evolution with meaning. There he develops the idea that "meaning is expressed through innate knowledge in the forms of organisms, the result of meaningful (i.e., coherent and sensitive to context) reading of the genetic text by the protolanguage of self-referential molecular networks, which have the same properties as written texts." The chapter is simultaneously lyrical and rich in biological detail. However, it invites a question: if there exists a compelling analogy, even an identity, between the cellular world and written texts, what reason is there not to consider the possibility that such form, where it first appeared, had the same cause as written texts, namely a mind?

[26]A book could be written connecting the focus on particularity and the denial of species in both late-medieval nominalism and Darwinism.

I look at the term species as one arbitrarily given, for the sake of convenience, to a set of individuals closely resembling each other, and that it does not essentially differ from the term variety, which is given to less distinct and more fluctuating forms. The term variety, again, in comparison with mere individual differences, is also applied arbitrarily, for convenience's sake.[27]

The matter, however, doesn't end there. Materialist reductionism does not only drain meaning from our zoological classifications; in its rejection of the living organism as real, it drains meaning from the very word *life,* damaging our language and thought, including the language and thought of scientists and science.

If we think about it, much of our language, any language, is rooted in the reality of organisms—in the belief that our words refer to living beings as "fundamental and irreducible units of life"—a point that can be easily understood if we recognize how many of our nouns refer to natural living things like cats, dogs, trees, pigs, human beings, cucumbers and so on. Of course, another significant portion of our nouns refer to nonliving things, both natural (like a stone or the Moon) and artificial (like a fence or a house). The nouns that do not refer directly to such things usually refer to parts or organs of such things (heart, hand); or to abstractions from things (like the antonym of abstract, *concrete*). Other nouns are more abstract still, like *incomprehensibility* or *ennui.* Verbs, of course, most often refer to the activities of objects (running, growing, barking, falling). Clearly, our language is rooted, in great part, in the reality of everyday things. This natural reliance of language on the reality and clarity of such things shows itself even more strongly as we dig into the etymology of words. Ralph Waldo Emerson summarizes the matter:

> Every word which is used to express a moral or intellectual fact, if traced to its root, is found to be borrowed from some material appearance. *Right* means *straight; wrong* means *twisted. Spirit* primarily means *wind; transgression,* the crossing of a *line; supercilious,* the *raising of the eyebrow.* We say the *heart* to express emotion, the *head* to denote thought; and *thought* and *emotion* are words borrowed from sensible things, and now appropriated to spiritual nature.[28]

---

[27]Charles Darwin, "Doubtful Species," chap. 2 of *The Origin of Species,* 6th ed. (New York: Mentor, 1958), p. 68.

[28]Ralph Waldo Emerson, "Nature," in *The Portable Emerson,* ed. Carl Bode and Malcolm Cowley (New York: Penguin, 1981), p. 19.

Further, words originally signify things in a complex way, rather than in some kind of simple-minded word-thing correspondence. As Emerson says elsewhere, "Every word was once a poem":

> For though the origin of most of our words is forgotten, each word was at first a stroke of genius, and obtained currency because for the moment it symbolized the world to the first speaker and to the hearer. The etymologist finds the deadest word to have been once a brilliant picture. Language is fossil poetry. As the limestone of the continent consists of infinite masses of the shells of animalcules, so language is made up of images or tropes, which now, in their secondary use, have long ceased to remind us of their poetic origin.[29]

It's important to keep all of this in mind when trying to grasp the far-reaching significance of the reductionist program, for if all is but a bump and grind of subatomic relationships, then the ontological status of the very beings of our experience—and, consequently, the language that is largely based on them—are undermined. If organisms are not real—if cats, dogs, trees and humans are really the accidental accretion of genotypic traits—then the substantive nouns referring to them are merely human constructs, including the "human" "doing" the "constructing." There's a reason postmodern author John Barth named one of his most famous stories "Lost in the Funhouse." Contemporary reductionist thinking tells us that everyday language, everyday experience, is a hall of mirrors, self-referential and vacuous, disconnected from reality.

This is not just a problem in regard to the reality of living things. The same would be true of almost all nonliving things large enough to be visible, since they too would be reduced to their atomic constituents. To a thoroughgoing reductionism, things like rocks and rivers are merely human referents to aggregates of matter, the visible forms of which have no ontological status.

But that isn't all that the acids of reductionism eat through. If both the living and nonliving things around us have no substantial existence, then the verbs and adjectives and abstractions, as well as all of the "fossil poetry" built from them, likewise lose their traction with reality. Grammatical structure assumes that organisms are "fundamental and irreducible units," wholes with their own ontological status. There are things out there to which the nouns *sheep* and *cow* refer. These things, existing as unified beings, can be the subject of unified self-directing acts—acts not reducible to genetic

---

[29]Ralph Waldo Emerson, "The Poet," in *The Portable Emerson*, ed. Carl Bode and Malcolm Cowley (New York: Penguin, 1981), pp. 252-53.

epiphenomena—so that ascribing action to them through verbs (i.e., predication) is to mean something about our world.

The ability to predicate meaningfully about real things, using a noun and a verb, lies at the heart of language. This includes simple statements like *the cow is* or more complex assertions like *the cow is chewing, the cow is running* or *the cow is lowing*. As unified subjects, cows and sheep also have real qualities (e.g., being brown, black, white, spotted, hot, old and hungry), so that such adjectives also mean something about something real. These actions and qualities can vary without the variations leading to a blurring of all boundaries and a reductionist loss of identity, so that we may truly describe such differences by adjectives and adverbs, such as *big* dog, chewing *loudly, very* brown, *pretty* old.

We could run through the other parts of speech, but the reader by now gets the point: in no small degree, both the meanings of language and its grammatical structure assume the reality of the organism. If that reality is denied, then the language on which it is based and the grammar by which it is structured become meaningless in the sense of their failing to illuminate, and to involve themselves with, being. If the materialist account is true, then meaning is a mirage and the deconstructionist, the nihilist, is our greatest sage.

This is not a private interpretation of matters but rather the history of what happens when the dogma of reductionism is accepted. It is played out in the linguistic turn in twentieth-century philosophy and in the epistemological crisis seeping into poetry and fiction and theater. It is philosopher Jacques Derrida's endless play of signs without grounded signifiers. The deconstructionists are just the latest to try and work out the implications of a world that is meaningless because it is pointless and pointless because it was randomly cobbled together.

The crisis of meaning reaches into every area of human inquiry, including the sciences. Many metaphors taken from the level of our ordinary experience, and those used in the service of science to make intelligible the intricacies of nature below or above the level of the everyday drama, would also be ultimately groundless. Indeed, there is a largely unnoticed admission of the uselessness of everyday speech in much of modern science insofar as scientific descriptions tend ever more completely to be expressed in terms of mathematics (and this is especially true among Darwinists insofar as they have given up studying actual living organisms and instead perform mathematical simulations of population and genotypic transformations, and computer simulations of alleged morphogenesis).

But even mathematics cannot escape from the acids of this reduction-ism. We take for granted the existence of mathematics, in all its permuta-tions, as a distinct mode of knowledge with its own set of intellectual tools that requires no explanation as to origin or effectiveness, and hap-pily use it as a means for understanding the order in nature. Yet as soon as we begin to inquire into its origin and effectiveness, we discover some obvious but important truths: mathematicals do not exist as objects like cats and dogs; nor do they exist in our minds from birth. Rather, they are learned. Further, they are learned not as abstractions that exist in their own right but, like our everyday language, through abstraction from ac-tual things.

There is an amusing philosopher's quip: "I've been to a lot of parties, but I've never met a number 7." The point is that numbers are abstractions, not entities you can see and meet. The number 7 that we write down is merely a symbol for the very abstract notion of seven-ness. The same can be said of geometrical figures, like circles, squares and so on. As used in geometry, they are abstractions purified of the substance, contingency and materiality of sensible objects.[30] Nobody can draw a perfect circle. And if nobody can draw a perfect circle, then nobody can draw a circle. The phrase *perfect cir-cle* is a redundancy; properly understood, a perfect circle is the only kind of circle. A geometric circle is, by definition, perfectly round. Anything short of perfection and it isn't a circle; it's just a pretender, a handy but less-than-perfect representation. Nature, of course, is three-dimensional and so un-friendly for a very different reason to the order of perfect circles. That said, we do not even find real, perfect spheres in nature. Instead we find sphere-like objects—like the moon, the earth, an orange, a ball—from which we abstract the notion of a sphere.

But even though we don't find numbers and perfect geometrical figures in nature, we needed things in nature from which we could abstract the mathematical and geometrical concepts to begin with. We are not born with abstract mathematical concepts imbedded in our minds; we learn them, and the foundation from which these concepts are learned is (as anyone knows who has taught mathematics to children) the everyday experience of ordi-

---

[30]That is, in abstracting the purely intellectual notion of sphere, we "leave behind" the sub-stance of what it was abstracted from, so that the notion of sphere may be abstracted from an orange, the Moon or a water droplet. Likewise, we don't include in the definition of sphere the contingency of the orange (that it has grown, that it will rot) or the particular materiality (the chemical constituents and internal structure).

nary unified objects—the reality which reductionism denies. Recall the
mathematician R. W. Hamming's words:

> I have tried, with little success, to get some of my friends to understand my
> amazement that the abstraction of integers for counting is both possible and
> useful. Is it not remarkable that 6 sheep plus 7 sheep make 13 sheep; that 6
> stones plus 7 stones make 13 stones? Is it not a miracle that the universe is so
> constructed that such a simple abstraction as a number is possible? To me this
> is one of the strongest examples of the unreasonable effectiveness of mathe-
> matics. Indeed, I find it both strange and unexplainable.[31]

To count sheep is possible precisely because sheep are each *unified* beings that bear a close *natural resemblance* to each other, a resemblance that
we take to be real. At its origin, mathematics demands both aspects: the
unity of the individual and the similarity within the group. To understand
this, imagine a world of disorder and dissimilarity, so that no distinct, unified
beings or things existed. In such a world (setting aside for the moment that
you wouldn't be there), you would never abstract the concept of "oneness,"
a concept based on the existence of distinct things that can be considered
"one" by being an obvious unified whole, such as, for example, one sheep,
one rabbit, one rock or one pigeon.

To state it somewhat whimsically but quite accurately, the ability to count
depends on the existence of sheep, that is, the existence of individuals be-
longing to obvious, natural kinds or species. The interesting thing is that humans abstract from the variations that occur among individual sheep and
concentrate on each sheep as a representative of what is common to the
species. From this commonness rooted in the species, we abstract the notion
of unity: one sheep, or if we might bend grammar to make a point, *one of*
sheep, one of this kind of thing we call sheep. And so, in all such instances,
the reality of the counting depends on the reality of the species.

From this, we note that in learning mathematics, humans naturally go
against the grain of Darwin's argument. Strange though it may seem to neo-
Darwinists, Darwin's assumption that the terms *species* and *variety* are
merely given for convenience's sake is part of a larger materialist and reductionist program that undercuts the natural foundation of counting and hence

---

[31]R. W. Hamming, "The Unreasonable Effectiveness of Mathematics," *American Mathematics
Monthly* 87 (February 1980): 84. Hamming is offering his argument as a further reflection on
physicist Eugene Wigner's famous essay, "The Unreasonable Effectiveness of Mathematics in
the Natural Sciences," originally published in *Communications in Pure and Applied Mathematics* 13, no. 1 (February 1960).

distorts the natural origin of mathematics. To put it more bluntly, in assuming that "species" are not real, Darwinism and the larger reductionist program burn away the original ties that bound the meaning of mathematics to the world and instead leave it stranded on a solipsistic island of the human imagination.[32] Mathematics becomes a set of abstractions without any real, original and natural foundation. Reductionism, then, does not allow us even to count the dead with any conviction.

From these theoretical ruminations, we move to the obvious and practical. We act as if living things are real, and not merely for convenience sake but because we know it to be true. Our very nature demands it in the practicalities of everyday life. With the phantom acid of Darwinist and materialist reductionism spilling all around us, we go right on counting sheep and goats and coins and roses. We treat ourselves and other living creatures as alive. We speak about cats and cows with perfect and practical confidence. In doing so, even through the darkness of imposed cultural nihilism, we have been heralding a return to common sense, spurred along by the cutting edge of science. In the midst of the reductionists' work to dissolve everything into bare atoms chasing after the wind, in the midst of the growing mist and darkness, an amazing thing has arisen in the West—so amazing that it borders on the mystical. The deep-down intelligibility of nature, of the cosmos, is reasserting itself ever more strongly and insistently, and scientists are the ones most filled with wonder (wonder mixed with a kind of gratitude that borders on religious awe) at the continual unveiling of its beauties. Mathematicians are ending up as mystics. Something has gone dreadfully right.

Albert Einstein wrote, "I have deep faith that the principles of the universe will be both beautiful and simple." But he found it strange that it should be so: "The most incomprehensible thing about the universe," he said, "is that it is comprehensible." Theoretical physicist Paul Davies is also struck by this. "Most scientists take it for granted that the Earth is both ordered and intelligible. And, the intelligible part, I feel, is really quite extraordinary." Davies is no theist and even says he's uncomfortable with the idea of miracles. Nevertheless, he continues, "It's one thing to accept the universe

---

[32]Reductionism also dismantles the human mind and imagination. In *The Mind's Eye* (New York: Basic, 1981), Douglas Hofstadler and Daniel Dennett conjecture that mind is merely an epiphenomenon of a certain level of neural complexity; thus it is really only its many millions of neural firings, themselves reducible to cells, chemistry, and the flux of matter and energy. It's worth considering what this would mean for the argument of their book, which, on their terms, is ultimately but a long display of neuronal fireworks.

as ordered; but ordered in a way that human beings are capable of understanding is an extraordinary thing."[33]

Hamming, then, is not alone in his amazement before the "unreasonable
effectiveness of mathematics." Nor is he alone in finding it "both strange and
unexplainable" that numbers and the cathedral of mathematics built from
them should apply so effectively at so many diverse levels of reality, both
above and below our proper scale of vision. This amazement is, we might
say, "Darwin-proof." Darwinism cannot reach back to the "first spinning
place," where the order of nature first manifested the mathematical principles we have only now discovered. The universe was supremely fine-tuned
and fit for biology long before biological survival of the fittest could possibly
work; and the strange, multilayered fitness of nature for mathematical analysis was present long, long before merely human mathematicians were there
to uncover it. Why, then, should we suffer any longer under the reign of
materialism and its subspecies, Darwinism?

There is no reason, except an all-too-human intransigence in the face of
mounting evidence. Science has always proceeded not by the deductive
proofs of logic, but by a competition among hypotheses in the uncompro-
mising realm of observation, where the very human explanations that best
account for the data are continually tested against reality. This only works if
hypotheses are humbly worn like a pair of glasses. Select a hypothesis using
a reasonable guess from the data, and if it seems to reveal more of the living,
breathing world, then the hypothesis has merit. If, after a time, it seems to
obscure more than does another pair of glasses, then it should be humbly
set aside. However helpful materialism has been in seeing certain material
aspects of nature, it has outlived its usefulness and obscures far more than
it clarifies.

Many reductionist-materialists are, of course, fond of claiming that their
methodology is historically responsible for the scientific revolution, but this
is incorrect. The birth of science required a rich ancestry of ideas and cir-
cumstances. In large part, the faith that drove many early scientists was the
prior commitment to cosmic orderliness and elegance inspired by faith in
a supreme artificer. That prior commitment is not difficult to uncover
among the founders of modern science. For instance, in his search for an

---

[33]This paragraph and the Davies quotation follow a section from the science film *The Privileged
Planet: The Search for Purpose in the Universe* (La Habra, Calif.: Illustra Media, 2004). The
screenplay for the film was cowritten by Jonathan Witt.

elegant description of our solar system, Nicholas Copernicus explained that he was motivated to uncover "the mechanism of the universe, wrought for us by a supremely good and orderly creator." It was a cosmos, he said, that "the best and most orderly artist of all framed for our sake." It was this that led to an expectation that the many and messy details of our solar system possessed some underlying unity according to which all those details would be fully intelligible. In short, Copernicus began with an assumption which contemporary science is once again rediscovering—that nature is ingeniously ordered.

We have referred frequently to "materialists." We have in view here not the average person plagued by doubts about questions of religion, or the experimental scientist attempting to search out new regularities in the order of nature, but rather the militant materialist, the village atheist, bent on scouring all evidence of the supernatural from the world. Such militant materialists have probably never made up more than 10 percent of the American population or much more than that in Europe. However, they have assiduously sought out and controlled entrance to the seats of academic power, and so they are represented in the tenured offices of higher education and the benches of our courts in disproportionate numbers. Against their efforts, the order and meaning of nature have reasserted themselves. Indeed, far from demonstrating the unreality of the organism, the materialist-reductionist program has instead laid bare how the organism embodies qualities common to the works of human genius—the qualities of depth and clarity, harmony and elegance, which are exquisitely fitted for a human audience and surprising the studious viewer at every turn.

Because the chemical elements themselves seem orchestrated, at least in part, for biology, and because organisms are not merely amalgams cobbled together but existing, acting, living wholes, it turns out that the everyday world in which we find ourselves is both a scientific beginning point and a scientific ending point. Biology cannot be reduced to chemistry and thence to physics. On the contrary, physics and chemistry have an end, a culmination, in biology; and biology is ultimately defined by the study of living things, especially those at maximal complexity, those of our everyday experience. The "reading" of nature's text begins there and returns again, where the most complex and delicate biological conditions are necessary if that most complex and extraordinary of biological forms on Earth—the human being—is to engage in science.

The collapse of reductionism, therefore, restores science. The true goal of science, contra reductionism, is not to explain away the everyday world of our experience, but to explain it. In doing so, a scientist is demonstrating that nature, like the work of Shakespeare, has both clarity and depth and that neither is an illusion. Reductionism grasped a piece of this truth, but held it so tightly and exclusively that even this piece became twisted and misshapen. The piece of truth, that nature has depth, was deformed into the corrosive and distorted dogma that the depths of mechanism were all there were and that our world of everyday experience itself was unreal, a mere epiphenomenon of the only real world, the subatomic world. In reducing reality to that single level, ultimately reductionists actually denied depth, flattening our world in a way that a flat-Earth zealot could only envy.

Against this, we may now insist on the commonsense view. Common sense, of course, is sometimes dazzlingly wrong, but it doesn't follow from this that commonsense conclusions are necessarily wrong. They're usually right or there would be none left to tell the tale, no survivors of a race of impractical idiots. In the present case, common sense is confirmed by the discoveries of the last century: macroscopic life is real, just as real as the depths of microscopic complexity from which it is built, whose ingenious forms make scientific discovery possible. In the final chapter we will consider where the sense of these discoveries leads.

10

THE END OF THE MATTER

A Meaningful World

The world is charged with the grandeur of God.
It will flame out, like shining from shook foil.

Gerard Manley Hopkins, "God's Grandeur"

AND SO WE COME FULL CIRCLE IN OUR STUDY OF GENIUS—a concept which grew from the Renaissance notion of the divinely inspired human artist imitating the creative potentialities of the Divine Artist and which has roots that go still further back, to the ancient Greek and Roman belief that the artisan, at his or her best, imitates the genius of nature.

The universal acid of materialist reductionism sought to deny all this; but now nature's genius has reasserted itself in the exquisitely functional and interdependent city of the cell, in the larger civilization of the multicellular organism and in the extraordinary fine tuning of the laws and constants of physics and chemistry for life and discovery. It is a realm of many cities and of cities within cities. None are too small to amaze.

And in rediscovering the full and astonishing drama of nature, we rediscover our very selves. We restore meaning to our strivings, our love of beauty, our joy in discovery. We regain our true depth and complexity. Indeed, the bare fact that we alone of the animals naturally desire to plumb nature's depths contradicts the reductionist assumptions that would divide us from ourselves. Our peculiarly human appreciation of beauty, for example, is as real as rocks and is far more profound than any Darwinist can allow; for it is not just an appreciation of surface form and color, but a deep affinity for harmony and elegance, stretching from the evident, real and sensible beauty of a flower or an animal to the rational understanding of the underlying complexity of each.

Almost all who study nature, in moments when they are unburdened by reductionist dogma, understand that this full range of elegant order is the highest object of science (and that the object is not just the discovery of some fortuitous chemical formula). A botanist, for example, is naturally drawn to study flowers by their beauty, a beauty residing both in their visible forms and in the intricate microscopic structures and processes that drive the flowers—a truth that comes out quite naturally in the way the botanist speaks when he isn't unnaturally worried about offending the guardians of materialist dogma. The more deeply a botanist delves into the rich and endless profusions of the plant world, the more he appreciates every growing thing, no matter how mundane or magnificent a particular plant may appear to the amateur. He treats the flower as any sane person treats the works of William Shakespeare, as a dramatic masterpiece of great depth.

Time and again, against the notion that nature is randomly ordered and hence ultimately meaningless and unintelligible, we find beauty, intelligibility and being entwined throughout nature. *Entwining* is not quite the right word, however, for these aspects really aren't separate to begin with, though we can mentally separate them, in abstraction. The elegant beauty of a rose, for example, isn't something added to the plant; the rose *is* beautiful in the fullest sense, from surface to microscopic depth. The beauty, and the beautiful intelligibility, of the flower—its identifiable form and color, its layers of integrated cellular and molecular structure, its ultimate chemical constitution—is not something added to or extrinsic to the flower; it *is* the flower. And the is-ness, or being, of the rose is meaningful to us insofar as we know it, to whatever degree we have penetrated its intelligible order, layer by layer, and understand how it is that the underlying layers of complexity culminate in a rose.

In other words, a rose is most meaningful to us when we understand it as a kind of dramatic culmination, one possible only because all these layers of complexity are integrated by and toward the whole, brought into harmony in and by the living form itself. Understanding how the elements or parts are brought together harmoniously in the whole is a central goal of science to which the analysis of the whole to its parts is a mere handmaid.

It should now be clear that meaning is inherent in nature, indicated by the fact that meaning in science is entirely dependent on nature. We can grasp this essential point by understanding the negative. If we don't know what a particular part of a plant is, then a crucial part of its meaning has eluded us. To add a name to it—"this is a sieve plate" or "this is a parenchyma cell" or "this is a stoma"—does not mean anything to us until we know what each part does and how it functions in relation to the overall life of the plant.

For example, consider the following description:

> A *stoma* is a specialized passage in the leaves of plants that allows carbon dioxide to pass into plants for the sake of photosynthesis and also allows the byproduct of photosynthesis, oxygen, to be released from the plant by diffusion. Stomata also release water in transpiration, a process that helps pull water from the roots up throughout the entire plant. Without stomata, plants would have to rely solely on atmospheric pressure to push water upwards and hence could only grow, at best, to less than a meter high.

Note that in order to grasp the meaning of the part, we comprehend it as a part of the whole plant; that is, we understand what it is by its purpose in the complex living unity of the whole plant, the entire being of the living plant, be it a redwood or a rose. Since every layer of complexity contributes to the living rose and these layers are knowable, the plant is truly meaningful to those who know its many layers.

Nor is the rose that the lover gives his beloved some late perturbation of matter resulting from "a chain of accidents reaching back to the first three minutes." Instead, the layers of complexity that make the rose possible include the original fine-tuning of the physical constants that allowed for the formation of galaxies and the many elements; the formation of our own solar system with its life-giving store of minerals; and the fine-tuning (cosmic and local) that allows for carbon, oxygen, water, carbon dioxide and other life-essential elements and compounds to mingle (allowing, but not by themselves causing, the extraordinary integration of the biosphere). Finally, there is the biochemical form of the rose itself, which allows it to function as a living unity, and of the human, uniquely equipped among the animals to appreciate both its depths and its surface beauty. Steven Weinberg's "first three minutes" are not pointless; they point to and culminate in the rose and in the scientist, the poet and the lover who hold it dear.

This a surprising discovery, but then, nature is full of surprises. One of the greatest is that disanthropism assiduously applied has ended in reviving an anthropism that stretches back to the origin of modern science and, further still, to the origin of the universe. No amount of philosophical wrangling or dispute could have established the importance of humans in the universe more certainly than the dedicated attempt by so many scientists to cast our race beyond the pale of ultimate meaning and purpose.

Insofar as humans are a culmination of the physical universe, anthropism reestablishes a high significance to humanity. And here the argument isn't troubled by the mere possibility of alien races circling other suns. Our use of the term *human* doesn't require a parochial sense, one even limited to planet Earth. Contemporary astrobiology shows us that technological life is so unlikely on materialist presuppositions that we may be alone in the galaxy, alone even in the universe.[1] In other words, there's nothing the least bit unscientific about the notion that we very well may be the only advanced organisms in the universe. But if there are extraterrestrials who can speak and write and fashion works of genius with hands that can bleed, who with their mind's eye can pierce "the deep wood" of nature's "woven shade" and kneel in wonder before the Maker on the other side, then in the larger sense there are humans far away as well as near—creatures who live on a world, like ours, constrained by the myriad of conditions that make such advanced life possible.[2] While such creatures may exist only in our imagination, the real conditions that define the possibility of complex, intelligent life do not. We are a culmination of those conditions, a culmination of the cosmic order.[3]

Here again we meet with the strongest argument for a designing intelligence of the cosmos. It is not just that from start to finish, we find a multitude of "parts" so well designed in anticipation of their culmination in complex biological life; it is that (as we have seen especially in the chapters on the history of chemistry) the intelligibility of the order both condescends to our capacities and continually exceeds them. Each science proceeds by discovering both its own shortcomings and a more comprehensive and complex order that demands new theoretical and experimental forays into the unknown. Once there, scientists again and again find some fortuitous guidepost directing them toward deeper insights, ever further into the superabun-

---

[1] See Peter Ward and Donald Brownlee, *Rare Earth* (New York: Copernicus, 2000), and Guillermo Gonzalez and Jay Richards, *The Privileged Planet: How Our Place in the Cosmos Is Designed for Discovery* (New York: Regnery, 2004).

[2] The phrasing here is borrowed from William Butler Yeats's "Who Goes with Fergus," in *Selected Poems and Two Plays of William Butler Yeats,* ed. M. L. Rosenthal (New York: Macmillan, 1962), p. 15.

[3] This is heady stuff and we would do well to remind ourselves that we did not make that order. The order of the universe is prior to and independent of our attempts to understand it. That is why our theories must be tested against nature. We are not creators of order but, at best, discerners of order, simultaneously servants and stewards of that order—not only for our existence but for the perfection of our understanding. If we say to that order *non serviam,* then we consign ourselves to irrationality and self-destruction.

dance of meaning that is the cosmos. The universe, and our privileged place in it, proves not only meaningful; the cup of its meaning continually overflows into mystery and wonder. The universe is crafted to condescend to our capacities as teacher to student and to draw us patiently upward; and the superabundance of intelligibility is a sign that it was made by a mind that far exceeds the merely human.

We analyzed the notion of genius first by examining Shakespeare. This was done because scientific reductionism, even though it begins by turning against God, necessarily ends by turning on itself, so that all human efforts are atomized, reduced to physical or chemical causes, including the efforts of the poet or the scientist. It ends in, to use poststructuralist Roland Barthes's phrase, "the death of the author."[4] The result is that that the commonsense understanding of Shakespeare as a unified, thinking, acting, creative human being who wrote extraordinary dramas about other unified, thinking, acting human beings is itself destroyed. The result is the denial of all meaningful human activity—not just the writing of drama but the drama of our everyday lives, including the activity of science itself. We strove to recover the native grasp of human genius to be able to see again the exquisite craftsmanship of nature, especially human nature.

We moved from the nature of genius to evidence of genius in nature. And just as it was evidence of Shakespeare's genius that his ordering powers both condescend to the everyday level and continually exceed our attempts to delve into its rich depth, so too nature's both condescending to and continually exceeding our attempts to grasp its exquisite order is evidence that nature is the work of a designing genius.

In 1800, four years before his death, the great chemist Joseph Priestley confidently penned his *Doctrine of Phlogiston Established* almost exactly a quarter century *after* Antoine Lavoisier had decisively destroyed this doctrine. Priestley wrote in a letter in January 1800, "I feel perfectly confident of the ground I stand upon."[5] Confident or not, he was already defeated. Ironically, it was Priestley's many successes and the amount of time he spent studying nature under the phlogistian paradigm that provided Lavoisier with a wealth of very precise experiments by which he could overthrow Priestley's cherished theory. Yet it was Priestley's very successes that instilled in

---

[4]Roland Barthes, "The Death of the Author," in *Critical Theory Since Plato*, rev. ed., trans. Stephen Heath, ed. Hazard Adams (New York: Harcourt Brace Jovanovich, 1992), pp. 1130-33.
[5]Quoted in J. R. Partington, *A Short History of Chemistry*, 3rd ed. (New York: Dover, 1989), p. 121.

him a kind of intellectual blindness to the truths Lavoisier had so brilliantly revealed, a blindness that kept him from seeing that nature was both deeper and more elegant than he would allow. Materialism is the phlogistianism of our age, its adherents blinded by their many successes in using the mechanistic analogy to explain natural phenomena.

Such blindness is actually a form of the most ancient of sins, pride; and this pride attacks science at its core. At the heart of the scientific enterprise is humility, the awareness that human reason is *human*. It is the reason of a rational animal, an animal not endowed with omniscience, an animal that is only a potential knower; it must proceed slowly and cautiously in finding out about the world, using sensation, imagination and abstraction to form general ideas, and then, using sensation and reason again, it must check these generalizations against the world. The order of being, the order of nature in all its intricate depth, is there all along. It is that which, with all our theory making and remaking, our experiment casting and recasting, we are struggling to know. The danger arises when, again, we become more attached to our current generalizations and speculations—however effective they might be—than we are to reality, and embrace our own theoretical constructions, partial and sometimes misguided, with a kind of idolatrous fervor. There is nothing easier to understand than a golden calf of our own making, and few things more difficult to understand than a calf of flesh and blood.

Our bringing up idolatry here is not a mere metaphorical device; rather it strikes to the very heart of the problem. Idolatry at its deepest is the worship of something that is human-made. In demanding that the universe must conform to human reason, to our theory, to what is simplest and easiest for us to understand, we are refashioning the universe into an idol. The fundamental error of the Pythagoreans (to recall chapter four) is that they were so taken with their grasp of geometry, with the clarity and certainty of the demonstrations and the bare beauty of the figures, that they made the universe a merely mathematical thing so that all of reality would be wonderfully clear and certain to them. The fundamental error of the modern mechanists is of a piece with the Pythagorean error: they are so taken with their fresh insights into the machinelike aspects of nature that they command the whole universe to be a watch and humanity to be mere automata.

Modern materialist reductionism embraces both idolatries in a union of mathematization and mechanization. Probably the most significant figure embodying this union is the acknowledged father of modern philosophy,

René Descartes.[6] Descartes dreamed of making a universal science, a "universal wisdom," that would grasp everything because everything in the universe could be reduced to homogeneous material extension, completely describable in terms of geometry and acting according to purely mechanical laws.[7] Since geometry was clear and certain to us and the machines we make are likewise clearly and certainly known to us, Descartes enshrined as the famous epistemological "rule" of his method that "the things we conceive very clearly and distinctly are all true."[8]

Instead of assuming that human knowing must above all strive to conform to the actual complexities of the world, Descartes's rule subtly shifted the stated locus of truth from nature to the human mind. Of course, it was precisely by eliminating those complexities of nature through his mathematical-mechanistic reductionism that Descartes thought he was able to transfer truth and certainty from nature to human beings. This revolutionary shift is presented in its most condensed form in his *cogito ergo sum,* "I think, therefore I am." The human "I" becomes the beginning point, or as he says in his *Rules for the Direction of the Mind,* "Nothing can be known prior to the intellect itself, since the knowledge of all other things depends upon this, and not conversely."[9]

Descartes tried to retain theism in his system, but later philosophers dispensed with his theism and what remained was the belief that the human "I" alone is the source of truth. Coupled with a reductionist view of science and nature, this ended in nihilism, the belief that there is no intrinsic order in nature to know and that human beings merely spin out meaning in a meaningless cosmos.

In the latter half of the twentieth century, following in the well-worn ruts of nihilism, deconstructionist philosophers attacked the very idea that we can know anything at all. The most influential of these was Jacques Derrida,

---

[6]For a deeper analysis of René Descartes along these lines, see David Lachterman, *The Ethics of Geometry: A Genealogy of Modernity* (New York: Routledge, 1989); Jacob Klein, *Greek Mathematical Thought and the Origin of Algebra* (New York: Dover, 1992); Louis Dupré, *Passage to Modernity: An Essay in the Hermeneutics of Nature and Culture* (New Haven, Conn.: Yale University Press, 1993); and Richard Kennington, *On Modern Origins: Essays in Early Modern Philosophy* (Lanham, Md.: Lexington, 2004).

[7]The phrase is from René Descartes *Regulae ad Directionem Ingenii* rule 1.

[8]René Descartes *Discourse* 4.33; quotations taken from René Descartes, *Discourse on Method and Meditations on First Philosophy,* trans. Donald Cress (Indianapolis: Hackett, 1980).

[9]Descartes *Regulae ad Directionem Ingenii* rule 8. Quotations taken from René Descartes, *Philosophical Essays: Discourse on Method; Meditations; Rules for the Direction of the Mind,* trans. Laurence Lafleur (New York: Macmillan, 1964), p. 175.

who accomplished his "deconstruction" of meaning by routinely conflating Descartes's simplistic view of language and comprehension with the more nuanced view of language stretching from the early Christian writers at least through the Renaissance, the view we may call logocentric.

Logocentrism was premised on the assumptions that the world was made through the wisdom of the Son, the Divine Logos (*logos* is a Greek word meaning "word," "reason" and "thought") and that our words, reason and thought could grasp the order and wisdom of creation. Of course, identifying the logos as the Son is particular to Christianity; but the understanding that nature revealed the logos—the reason, the surpassing wisdom of its maker—was not. It was found, in one form or another, among many Greek and Roman philosophers, in particular the Stoics. Derrida thought he was deconstructing logocentrism; but really he was only attacking Descartes's faulty epistemology (which unfortunately still dominates our intellectual field of vision). Derrida critiqued Descartes's field of vision, neglecting the more profound understanding of reality entailed in authentic logocentricism.

The problem with Descartes's construction is not only intellectual pride, but, following on it, his ignoring the idea that sin has clouded our intellect. Michael Edwards explains the errors made by both Descartes and Derrida. Descartes's *cogito* (the "I think" in "I think, therefore I am") failed to take into account human fallibility: the many ways the human mind can be darkened in its understanding without realizing it, taking its own strengths and insights to be signs that it enjoys a godlike freedom from error and failing to see its own inherent weaknesses and sinful distortions. This is especially true of Descartes's attempt to identify what is true with what appears clear and certain to us. As for Derrida, his critique failed to take into account the difference between Descartes's view and the older, orthodox view that recognizes human intellectual fallibility and the great distance between human and divine intelligence.[10] As Edwards explains, "To deconstruct logocentrism" as Derrida understood and defined it "is to discover the fallacy not of Logos but of what our worldly metaphysics has made of it by proceeding as if there were no Fall."

The traditional Christian view differs markedly both from Descartes's view and from Derrida's nihilism, the latter of which denies that we can

---

[10]Michael Edwards, *Towards a Christian Poetics* (Grand Rapids: Eerdmans, 1984), pp. 221-22. "However clear the thought of the world may be to God," Edwards writes in orthodox Christian terms, "for us there can be no simple presence of a world, since . . . creation has fallen into 'vanity' and 'corruption.'"

know anything (other than Derrida's claim that we can't know anything). We are as Shakespeare presents us—creatures in between, comedic, tragic, romantic, striving and toiling in all things. And the human striving to know partakes of what the great mathematician and philosopher Blaise Pascal termed *grandeur* and *misère*. Essentially, Derrida charges the Western tradition with a view of language and knowledge that takes account only of *grandeur*, limpid and "blithely sinless," moving straightforwardly from one triumph to the next. Having shredded this Cartesianism, Derrida falls into the opposite error, setting up a view of language that is pure *misère*, unmitigated nihilistic darkness, a language of unmeaning fit for a meaningless world. In this, Derrida has inadvertently done us an invaluable service, Edwards explains, for he has traced out the implications for meaning in a world without God: by removing the Author, the materialists created a meaningless drama.

This is perhaps the central point of Derrida's work, a point that emerges only with difficulty amidst his tortured prose. Serving as interpreter, Edwards delineates the deconstructionist's picture of a world without a Maker, alternately paraphrasing and quoting from Derrida's *Of Grammatology*: "To remove God, the guardian of the 'transcendental signified' which guarantees that the difference between signified and signifier is 'somewhere absolute and irreducible,' is to cast the signified back towards the signifier, within a process of signifying from which there is no exit."[11] In such a universe, "Meaning, being, truth, no longer exist outside of the sign, before language and independently of it, 'within the full presence of an intuitive consciousness.'" Rather, as Derrida says, it is a world of "signs leading to signs in a desperate infinity of transfer, in which everything only manages to mean everything else . . . a contemporary version of hell, in a world unnamed and incapable of being re-named."[12] It is, in short, a world in which there is no "meaning in the data," but only data forever in search of meaning, forever unsatisfied.

Our journey through the arts and sciences suggests that such a view is pessimism instead of realism and that the scientist's arduous and sometimes misguided exploration of the world, while not one of pure *grandeur*, is by no means only the blank nihilism of *misère* either. To use against them one of the deconstructionists' own rhetorical weapons, the journey is not either

---

[11]Ibid.
[12]Ibid.

grandeur or *misère* but both/and, and that is a sign of its humanness. Scientists can move forward, having faith that there is order to be uncovered, but only if they recognize that their models and presuppositions may demand revision in light of that order.

That is a great and salutary lesson nature has condescended to give us. Standing before the cathedral of the world, we realize that we should not assume we comprehend all the designs of a genius so daunting, imagining that our own reasoning is the plumb line of the world. Humility, as much as boldness and intellect, is critical to scientific discovery, for it leads us to expect the mysterious—a mystery that is not darkness but a superabundance of light. Nor does this humility imply a readiness to give up, for the humble scientist also humbly refuses to pretend she knows what secret features of the physical world have been set beyond the reach of science, either momentarily or forever.

The term *humility* is frequently misunderstood. Anthony Bloom explains:

> The word 'humility' comes from the Latin word 'humus' which means fertile ground. To me, humility is not what we often make of it: the sheepish way of trying to imagine that we are the worst of all and trying to convince others that our artificial ways of behaving show that we are aware of that truth. Humility is the situation of the earth. The earth is always there, always taken for granted, never remembered, always trodden on by everyone, somewhere we cast and pour out all the refuse, all we don't need. It's there, silent and accepting everything and in a miraculous way making out of all the refuse new richness in spite of corruption, transforming corruption itself into a power of life and a new possibility of creativeness, open to the sunshine, open to the rain, ready to receive any seed we sow and capable of bringing thirtyfold, sixtyfold, a hundredfold out of every seed.[13]

Fruitful scientists keep their bare feet in the earth. They listen carefully rather than command, looking for ways for the order of nature to speak. This brings us to a quality of genius at work that we have not yet mentioned—patience. Albert Einstein once insisted that he was different only in sticking with a problem longer than anyone else. This was hyperbole, as was George-Louis Leclerc de Buffon's declaration that "patience is only genius"; but such remarks do remind us of genius's homeliest virtue.

And so, just like good and humble readers of Shakespeare, the most fruitful scientists assume that the text of nature goes deeper still, that reality has

---

[13]Anthony Bloom, *Beginning to Pray* (New York: Paulist, 1982), p. 35.

still more to reveal and hence more meaning than we are now able to express—in short, that the universe, and our world in particular, is meaning-full, containing a depth of complex, integrated order that continually overflows our merely human cup. Nor does that order flow out of and into the void of nihilism, signs without signifiers in an endless fall of transfer. No. These signs point to an artisan and teacher of consummate genius, the Genius of nature.

We offer one final consideration. In the first chapter, we warned readers that we were not going to focus on the great problem of evil but on the more fundamental problem of meaning and the loss of meaning. Again, we maintain that the problem of evil is secondary precisely because evil is parasitic on good, in the same way that disorder is parasitic on order and meaninglessness on meaning. All too many people go at it the other way around, arguing that the presence of evil in the world both makes it meaningless and proves that there is no creator. Darwin himself famously complained, "I cannot persuade myself that a beneficent and omnipotent God would have designedly created the Ichneumonidae [a type of parasitic wasp] with the express intention of their feeding within the bodies of Caterpillars."[14]

As with the position of nihilism, this too is self-defeating. As Richard Dawkins, an arch-Darwinist and self professed atheist, recognizes, in a cosmos that is the result of accident, there is no good or evil: "Nature is not cruel, only pitilessly indifferent. This is one of the hardest lessons for humans to learn. We cannot admit that things might be neither good nor evil, neither cruel nor kind, but simply callous—indifferent to all suffering, lacking all purpose."[15] To remove God and enthrone chance removes the reality of both good and evil. Therefore, it makes no sense to say "I can't believe in God because of the existence of evil," for the affirmation of atheism denies the reality of both good and evil.

Materialists might counter that the suffering would certainly be evil if the world really were a place of good and evil, and so the world as we experience it is inconsistent with the idea of a benevolent, omnipotent creator. They are asserting that, judged from within, theism is incoherent and, therefore, false.

If you are a materialist making this argument, be forewarned: it comes with a risk. For the complaint against evil to have any tread, you must at

---

[14]Quoted in Dawkins, *River out of Eden* (New York: Basic, 1995), p. 95.
[15]Ibid., pp. 95-96.

least temporarily enter the world of the theist, a world where good and evil are regarded as real—and that is the string that may lead you out of the cave of reductionism; for in quieter moments, face to face with good and evil, the human heart knows that good and evil are real and not apparent.

The reality of evil does not make it easy to understand; as we've seen, real things tend to be more difficult to fathom than human-made caricatures or idols. Thus theists, specifically Judeo-Christian theists, don't gaze at our wounded world through rose-colored theological glasses. Rather, they recognize that the mystery of evil is near the heart of the drama of their dogma—and that at the very heart is the even deeper mystery of goodness.

Such a view is fitting. If the world is the work of surpassing genius, as the evidence laid out in this book suggests, we should expect mystery. The point becomes clear from our consideration of human genius. If we spend any time appreciating the work of some great human artist, at some point we will feel overwhelmed by the profundity of the artist's work. This attitude of awe and wonder is good and proper, especially as we regard the genius of nature, for a traditional understanding of the elements of a work of genius insists on the qualities of paradox and mystery.

The greatest surprise, in a work as rich as our world, would be to find neither.

Index